DODD, MEAD

Wonders of the Mosquito World by Phil Ault
Wonders of the World of Bears by Bernadine Bailey
Wonders of Animal Migration by Jacquelyn Berrill
Wonders of Animal Nurseries by Jacquelyn Berrill
Wonders of the Monkey World by Jacquelyn Berrill
Wonders of the Arctic by Jacquelyn Berrill
Wonders of the Woods and Desert at Night by Jacquelyn Berrill
Wonders of the World of Wolves by Jacquelyn Berrill
Wonders of Alligators and Crocodiles by Wyatt Blassingame
Wonders of a Kelp Forest by Joseph E. Brown
Wonders of Rattlesnakes by G. Earl Chace
Wonders of the Pelican World by Joseph J. Cook and Ralph W. Schreiber
Wonders Inside You by Margaret Cosgrove
Wonders of the Tree World by Margaret Cosgrove
Wonders of Your Senses by Margaret Cosgrove
Wonders of Geese and Swans by Thomas D. Fegely
Wonders of Wild Ducks by Thomas D. Fegely
Wonders Beyond the Solar System by Rocco Feravolo
Wonders of Gravity by Rocco Feravolo
Wonders of Mathematics by Rocco Feravolo
Wonders of Sound by Rocco Feravolo
Wonders of the World of the Albatross by Harvey I. and Mildred L. Fisher
Wonders of Sponges by Morris K. Jacobson and Rosemary K. Pang
Wonders of the World of Shells by Morris K. Jacobson and William K. Emerson
Wonders of Magnets and Magnetism by Owen S. Lieberg
Wonders of Measurement by Owen S. Lieberg
Wonders of Animal Architecture by Sigmund A. Lavine
Wonders of the Bat World by Sigmund A. Lavine
Wonders of the Bison World by Sigmund A. Lavine and Vincent Scuro
Wonders of the Cactus World by Sigmund A. Lavine
Wonders of the Eagle World by Sigmund A. Lavine
Wonders of the Fly World by Sigmund A. Lavine
Wonders of the Hawk World by Sigmund A. Lavine
Wonders of Herbs by Sigmund A. Lavine
Wonders of the World of Horses by Sigmund A. Lavine and Brigid Casey
Wonders of the Owl World by Sigmund A. Lavine
Wonders of the Spider World by Sigmund A. Lavine
Wonders of the Dinosaur World by William H. Matthews III
Wonders of Fossils by William H. Matthews III
Wonders of Sand by Christie McFall
Wonders of Stones by Christie McFall
Wonders of Gems by Richard M. Pearl
Wonders of Rocks and Minerals by Richard M. Pearl
Wonders of Barnacles by Arnold Ross and William K. Emerson
Wonders of Sea Gulls by Elizabeth Anne and Ralph W. Schreiber
Wonders of Hummingbirds by Hilda Simon

A museum specimen of the horny or bath sponge, Hippospongia canaliculata

Wonders of Sponges

Morris K. Jacobson
and
Rosemary K. Pang

ILLUSTRATED WITH PHOTOGRAPHS

Dodd, Mead & Company · New York

ACKNOWLEDGMENTS

Illustrations on pages 16, 28, and 70 are from *Animals without Backbones, An Introduction to the Invertebrates*, 2nd ed., by Ralph Buchsbaum. Copyright 1938 and 1948 by The University of Chicago. Reprinted by permission of the publisher.

Illustrations on pages 14 (bottom) and 46 are from *The Invertebrates: Protozoa through Ctenophora* by Libbie Henrietta Hyman. Copyright 1940 by McGraw-Hill Book Company. Used with permission of McGraw-Hill Book Company.

Illustrations on pages 32 and 33 are from *Kurs Zoologii*, G.G. Abrikosova and L.B. Levinsona, eds., Moscow, 1955.

Illustration on page 56 is from *Commercial Fisheries Review*, Fish and Wildlife Service, U.S. Department of the Interior, Vol. 29, No. 10.

We gratefully acknowledge the help and guidance given us by our editor, Jennifer Anderson. We would also like to thank Dr. Henry M. Reiswig of McGill University (Canada) for providing the underwater photographs of sponges.

Picture Credits: American Museum of Natural History, pages 2 (frontispiece), 6, 14 (top), 18, 19, 20, 37, 53, 63, 66; M.K. Jacobson, pages 40 (left), 54, 58; R.K. Pang, page 38; H.M. Reiswig, pages 10, 11, 22, 25, 26, 31, 34, 40 (right), 41, 45, 48, 49, 50.

Copyright © 1976 by Morris K. Jacobson and Rosemary K. Pang
All rights reserved
No part of this book may be reproduced in any form
without permission in writing from the publisher
Printed in the United States of America

Library of Congress Cataloging in Publication Data
Jacobson, Morris K
    Wonders of sponges.

    Bibliography: p.
    Includes index.
    SUMMARY: Surveys the classification of different types of sponges, their distinguishing characteristics, life functions, and usefulness to man. Briefly discusses the sponge-fishing industry and sponge collecting.
    1. Sponges—Juvenile literature. [1. Sponges]
I. Pang, Rosemary K., joint author. II. Title.
QL371.J23   593'.4   75-38363
ISBN 0-396-07300-X

Contents

1. Introduction to Sponges — 7
2. The Classes of Sponges — 13
3. The Life of Sponges — 23
4. Curious Sponges — 36
5. Sponges and Man — 52
6. Working with Sponges — 64
   Glossary — 72
   Annotated Bibliography — 75
   Index — 77

A museum specimen of the horny or bath sponge, Euspongia zimocca

1
Introduction to Sponges

It may surprise you to be told that the sponge you use for your bath or which a painter uses in his work, can actually be a skeleton, the skeleton of a very interesting animal which lives in the sea. Although this sounds hard to believe, it is nevertheless true. Of course the skeleton isn't hard, white, and bony like the skeletons we are used to, but it is still a skeleton.

Most people take a sponge to be a kind of "thing" which can be bought in a drugstore or a hardware and paint store. Actually, it once used to be part of a living animal, which, like all animals, has to take in oxygen, eat and digest food, discharge waste products, protect itself against enemies, and reproduce its kind. In this book we intend to describe how the sponge as an animal performs these functions. We will also show that, strange as some of these ways may seem, they are basically the same as in all forms of life, even our own. We will also write about the uses mankind has made of sponges and how sponges provide other animals with food and shelter.

Since bath and industrial sponges are used by human beings, a fair-sized sponge fishing industry can be found in various warmer parts of the world. Unfortunately, by far the largest number of sponges are of very little, if any, use. Some are even harmful and dangerous, while many do not affect us at all.

Many sponges are found in the shallow waters of bays and oceans. They are particularly abundant in the waters of Florida, the West Indies, the Mediterranean and Red seas, Mexico, Japan, and California. Frequently after strong storms, sponges are thrown up on the beaches in large numbers, even as far north as Maine, Alaska, and Canada. Sponges are also surprisingly abundant in the icy seas of the Antarctic. They are found in all the oceanic regions of the world. They live at all depths, from shallow intertidal areas to as much as 21,000 feet (about 4 miles) below the surface. Some kinds of sponges are found only in fresh water. Today sponges are an exceedingly abundant group of animals, numbering about 10,000 different species, and fossil sponges have been found from very early times right up to the present. Thus, they are an important part of the Animal Kingdom.

Scientists refer to all sponges by the word *Porifera*, a very fitting name as we shall see. When a living bath sponge is brought out of the sea, it looks quite unlike the object which we buy in a store. It is smooth, nearly round in shape, and covered by a dark leathery skin with a few large holes in it. But when one looks very closely, one sees that the slimy exterior surface is perforated by many, many tiny pores, barely visible to the naked eye. All sponges have such pores. For this reason zoologists call sponges the *Porifera*, a word made up of the word *pore* and the Latin verb *ferre*, "to bear" or "carry."

Sponges have been known to mankind for a very long time. In fact, sponges are mentioned in both the *Iliad* and the *Odyssey*, and the famous Greek scholar and philosopher, Aristotle, wrote about the nature of sponges. Like most people, he was puzzled by sponges and didn't know whether to consider them plants or animals. Though he thought that they might be animals, he was puzzled by the fact that, like plants, all mature sponges are

An old print of a large Neptune's goblet, Poterion

sessile—that is, they are fixed to one place and cannot move from it. Finally he decided that this peculiar creature showed that it had characteristics of both plants and animals. Many naturalists agreed with him, but for centuries others thought of sponges as plants.

In 1765 a man by the name of John Ellis noted that the sponge could circulate water through itself and that the sponge was able to change the size of the hole through which the water flowed. Plants cannot usually do such things. In 1825 a Scottish naturalist named R. E. Grant also noticed the flow and described

Ageles, *the "elephant ear sponge." Each unit measures about 2½ x 10 inches.*

it as "a living fountain vomiting from the circular cavity an impetuous torrent of water." It was he who showed that sponges do not take in free chemicals from the water like plants, but eat particles of organic food, tiny plants and animals, as other animals do. This finally clinched the argument. Dr. R. E. Grant invented the name *Porifera* in 1836. In honor of his achievements, a genus of sponges has been called *Grantia*.

Sponges come in many different shapes and colors, and a variety of sizes. Some are tall with arching branches and some are merely flattened algaelike growths covering stones and other hard objects. Some are thin-walled tubes, some perfect spheres, others look like pinkish cakes (*Geodia*), and some like maroon-colored plates (*Petrosia*). Some are treelike, some bushy, and others simply shapeless. Some are huge, reaching three to six feet in height and shaped like elephant ears (*Ageles*), others almost as large look like flower baskets (*Verongia*). Some are shaped like goblets, and one kind is even called Neptune's goblet (*Poterion*). Another type of sponge is so big (*Xestospongia muta*) that a diver can sit comfortably in the central cavity and peep out. And other sponges, even when they are full grown, are no bigger than a bean. Sometimes the shape of a particular sponge species may be different in different environments. For example, a sponge may have one shape in calm, still water and another shape in very rough, fast-moving water. This often makes identification of the sponge species difficult.

The surfaces of sponges also differ. Some are smooth, even velvety, others are rough, and in some the surface is prickly like a pincushion. Sponges also vary in texture. Some, like the bath sponges of the family Spongidae, are strong but flexible,

Left: A huge specimen of the sponge, Verongia, *weighing about fifty pounds, with a capacity of five gallons. Right: The demosponge,* Xestospongia muta *(diameter: about 18 inches), with a grouper sleeping inside. Look closely!*

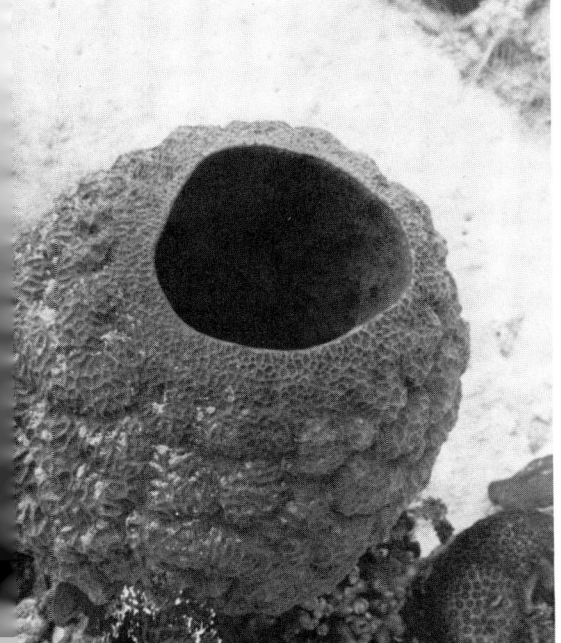

others are soft and mushy, and still others are hard and stone-like.

Sponges can be lemon yellow or brilliant red, vivid green to the most delicate blue and deeper violet, and every shading in between. The most colorful sponges of all are found on coral reefs in warm waters. Together with the corals themselves, the brilliant fish, and the gorgeous seashells, they form one of the most colorful combinations in all nature.

A very beautiful sponge (*Verongia lacunosa*) has a ridged outer surface, the tops of the ridges being yellow, and the valleys between a vivid green. Some sponges from the ocean depths are pure white. F. S. Russell and C. M. Yonge, in their book, *The Seas*, describe one of these deep-sea sponges as "dead white in color and of strange shapes like the whitened bones of some prehistoric monster." It is important to note that when a sponge is preserved in alcohol or dried in the sun, the color is usually lost. Even the most brilliant red and orange sponges look drab and rather uninteresting when they are preserved. However, there is one sponge called *Cliona schmidti*, which keeps its rich and striking purple color even when it is dried or preserved in alcohol. It is one of the very few members of the Porifera to have this characteristic.

2
The Classes of Sponges

Biologists who specialize in the study of sponges are called spongiologists. Like most experts, they have divided the animals they study into several large, related groups called classes. It will help us to understand the world of sponges better if we examine the classes into which the phylum Porifera is divided. But before we do that, we should discuss certain characteristics which are important in classifying sponges.

SPICULES

Spicules comes from the Latin word *spiculum*, which means "a point" or "a dart." These structures, found in many sponges, are usually very tiny. Many can be seen only under high magnification. Spicules may be either glassy (*siliceous*) or limy (*calcareous*). Basically they consist of a bar which can be smooth or spiny, straight or bent in the middle, and rounded, pointed, or knobbed at the ends. Sometimes many bars come together and form spicules of strange and beautiful designs. Sometimes the spicules have arms like stars, reaching out in different directions. Some spicules have odd shapes. They may look like tiny cuff-buttons, anchors, horseshoes, S- and C-shaped

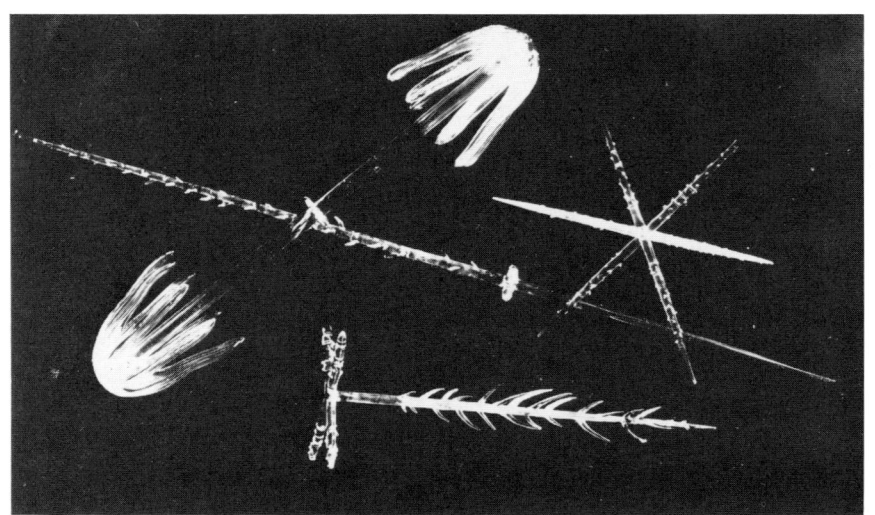

Above: Microscope picture of spicules. Below: Spicules of the glass sponges Hexactinellida

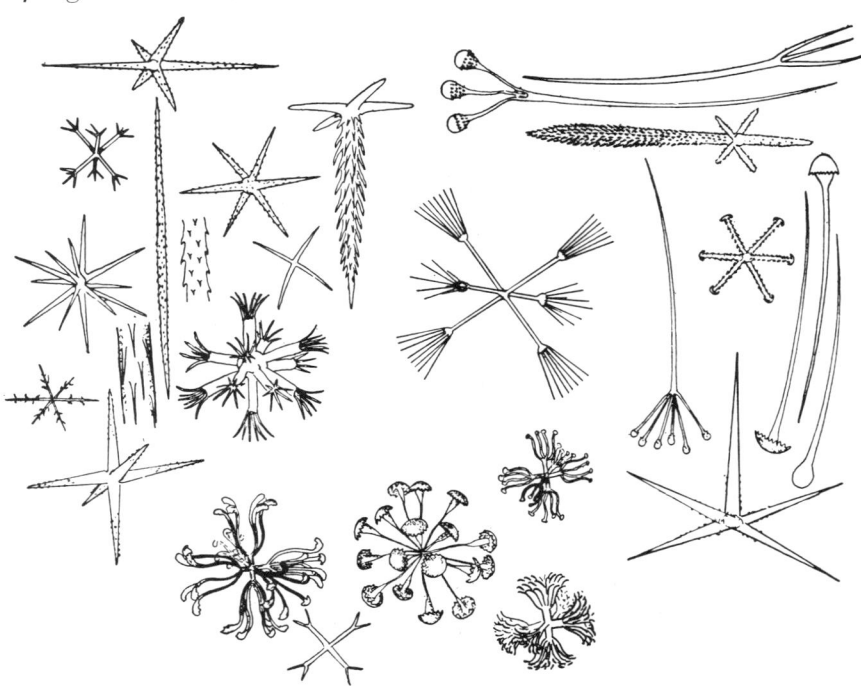

hooks, or coiled springs. What a pity that most of them aren't easier to see!

Since spicules are so important in determining in which class —and also in which species—the sponges are placed, spongiologists use a great many technical and specialized terms to describe them, far too many to discuss in this book. However, we should learn at least two. We have already seen that spicules can be either siliceous (glassy) or calcareous (chalky). They can also be very tiny or somewhat larger. The tiny spicules are called *microscleres*, from the Greek words *mikros* meaning "tiny" and *skleros*, which means "hard." The larger ones are called *megascleres*, from the Greek words *megas* meaning "big" and *skleros*. These two words are very important in the study of the Porifera.

The microscleres have little to do with the formation of the skeleton. In fact, spongiologists are not quite sure what their function really is. But they are not at all puzzled about the megascleres. They may be scattered throughout the sponge body or they may be bound together into long, fiberlike bundles which cross each other and interlock in all directions like the hairs that make up felt cloth. It is this type of structure, made up of megascleres, which forms the skeletons of some sponges. Sometimes grains of sand or bits of broken shells enter the body of the sponge and these too become part of the skeleton.

Not all sponges have both kinds of spicules. In one class of sponges, as we shall see directly, the spicules are all more or less alike in size. In most sponges which have spicules, however, both megascleres and microscleres are found. They appear in many different shapes and forms, and they are either glassy or limy. These shapes and forms are important if one wants to learn the Latin or scientific name by which a sponge species is known.

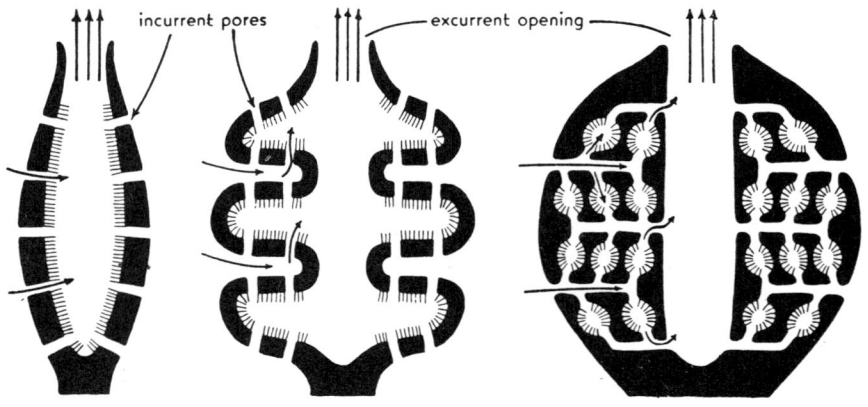

Types of sponge bodies (stylized): Left, simple or asconoid type; Center, more advanced or syconoid type; Right, complex or leuconoid type

TYPES OF SPONGE BODIES

Another important characteristic which spongiologists use to classify the Porifera is the type of sponge body. There are three types, as we shall now see.

The simplest and most primitive type of sponge consists of a saclike or tube-shaped body. The walls are thin and they have short, straight, and nonbranching channels leading from the outside into the interior. This type of body is called the *asconoid* type, from the Greek word *askos* meaning "a bladder," which in a way it does resemble. The same word is also found in the name of the genus *Ascon* which has the simple "asconoid" body type.

In other sponges the walls remain thin but they begin to fold inward at regular intervals and form little fingerlike extensions. Thus the walls are more complicated than those of the simple asconoid types. Therefore, spongiologists believe that this type of body is more advanced. They call it the *syconoid* type, from the Greek word *sykos* meaning "a fig." There is indeed some resemblance between a syconoid sponge and a large, wrinkled, ripe fig. The word *sykos* also appears in the name of the genus *Sycon*, whose members have a syconoid body.

The third type of body has an even more complicated kind of wall. It is very thick and full of tremendously twisting and branching tunnels and channels as can be seen in the well-known commercial sponges. For this reason it is considered to be the most highly advanced of all the sponge types. It is called the *leuconoid* type. The sponges of the genus *Leuconia*, among many others, have this type of body. *Leuconia* and *leuconoid* both come from the Greek word *leukos* meaning "white." Unfortunately this is not of much help in describing the leuconoid body type.

Of the three body types, the vast majority of sponges, when mature, have the leuconoid type of body. It is by far the most popular body type among the Porifera. On the other hand, relatively few sponges keep the simple asconoid body type all their lives, and only a few more have a syconoid body.

THE DEMOSPONGES

The largest and by far the most important class of sponges is called *Demospongiae*. This word comes from the Greek *demos* meaning "the ordinary or common people." This name was selected because all the "common" or "ordinary" sponges—the bath and commercial sponges—belong to this class. All sponges like these have a skeleton made up of a material known as *spongin*, the common, spongy material which we all know so well. Spongin is a very interesting material. Chemically it resembles silk and, like silk, belongs to a group of proteins called *scleroproteins* or hard proteins. (The Greek word *skleros* means "hard.") Other familiar objects consisting of scleroproteins are hair, fingernails, and horn. (Indeed, the bath sponges are often referred to as "horny sponges.")

But if all sponges containing spongin are demosponges, the

A *museum specimen of the red beard sponge (*Microciona*), which has been restored to look as it did in life*

opposite is not true. Not all demosponges contain spongin. There are a great many without spongin which also belong to this class. Some demosponges, instead of spongin skeletons, have a vast number of the tiny spicules to serve as a stiffening for their shapes. Many have both spongin and spicules, and some have no real skeleton at all. As a matter of fact, the Demospongiae is the biggest single class of the Porifera, including more than four-fifths of the sponge species known. It is also a very puzzling assemblage of creatures. However, they have certain features in common by which most can be recognized.

When they are mature, the demosponges have a leuconoid type of body with well-formed walls and an intricate system of channels and tunnels. When they have spicules, these can be both megascleres and microscleres, and they are always glassy (siliceous), never limy (calcareous). Many demosponges have skeletons containing or consisting of spongin, but some have no skeleton at all. Using these as guides, the reader should be able to recognize most demosponges he is likely to meet.

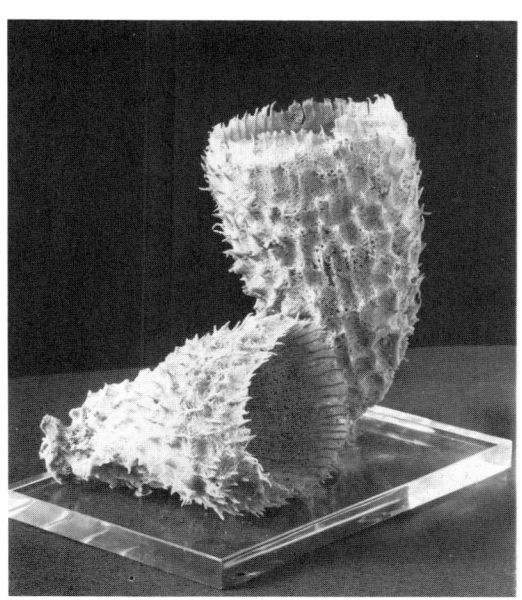

A beautiful museum specimen of the demosponge, Stelospongia, from the Bahamas, West Indies

THE GLASS SPONGES

There are three other classes of sponges. The most beautiful of all sponges, and some of the most beautiful structures found in nature, are the glass sponges. They look as unlike the ordinary bath sponge as one could possibly imagine. They clearly belong in their own class of the Porifera.

The glass sponges are found in deep water. Most live between depths of 1,200 to 2,700 feet, but some are even found as deep as three miles below the surface of the sea. They are particularly abundant in the West Indies and the eastern Pacific, from Japan southward to the East Indies.

One of the most beautiful glass sponges is Venus's flower basket. It comes from very deep water off the coasts of China, Japan, and elsewhere in the warm parts of the Pacific Ocean. It is about eight inches high and two or three inches wide and forms a beautiful upright column composed of a network or lattice of glassy, closely interwoven threads. When the sponge is cleaned, it looks as though it is made of the most beautiful

Left: A museum specimen of the glass rope sponge, Hyalonema sieboldii. *Note the bundle of very long root spicules. Right: A museum specimen of the small calcareous sponge,* Grantia, *growing on a muscle shell*

spun glass. Its scientific name is *Euplectella*.

The class to which the glass sponges belong is called *Hexactinellida*. This is composed of the Greek words *hexa* for "six" and *aktinos* meaning "a ray." This name is based on the fact that the spicules of these sponges are the only ones which have six glassy arms or rays. But these are not simple rays. They develop into very complicated and beautiful structures. Some rays end in mushroomlike caps, some are twisty and widened like flower petals, others have ends like tiny whisk brooms, and some bushy, featherlike plumes. Under the microscope some of these spicules are as beautiful as snowflakes.

Hexactinellida also produce another kind of spicule. Some

glass sponges live in very deep water on a bottom of mud and ooze. To keep from being washed away, they have to be securely anchored. Even more important, they must be kept above the mud. Therefore, the anchor of these sponges consists of a bundle of very large glassy spicules (the very largest in the entire phylum Porifera), raising them above the bottom ooze. Thus, though most spicules can be seen only under a microscope, here we have one kind which can be seen easily with the naked eye. But a microscope is still needed to be able to see the beautiful structures which ornament the sides of these huge spicules.

Chalky and hard sponges

There are two more classes of the Porifera which also have interesting features. One of these is the class named *Calcarea*. This is a good name for this class because the spicules, instead of being made of glassy material as in the Demospongiae and Hexactinellida, are dull and chalky, or calcareous. Moreover, these spicules are all more or less the same size and thus are not divided into microscleres and megascleres. The Calcarea are all quite small, some no bigger than a bean, and others grow to about five or six inches. Many are shaped like small vases with a simple asconoid body type, but others have a syconoid or leuconoid body type. The entire surface has tiny bristles sticking out all over. These bristles are the pointy tips of the chalky spicules. The Calcarea are found in shallow water even in colder climates, and many can be collected from the blades of sea grasses they cling to. Two kinds, one an asconoid sponge called *Leucosolenia* and another called *Sycon* (or *Scapha*), a syconoid sponge, are often studied in biology classes.

The last class of sponges is called *Sclerospongiae*, from the word *skleros*, which as we already know means "hard" in Greek. These sponges are composed of so much hard material that for years they were taken to be parts of the coral formations. Although the other three classes of sponges were established many years ago, the recognition of this fourth class has come about only in recent years. Dr. W. D. Hartman and Dr. T. F. Goreau set up the class to include some curious creatures found on reefs off the northern coast of Jamaica which did not seem to fit into any of the older three classes of sponges.

The Sclerospongiae produce glassy spicules, which are often hidden under a large mass of calcium, also produced by the sponge. They are found on coral reefs in the tropics, mainly in Jamaica, where they do not live in the open, in bright sunlight. They prefer to live in shaded cavities formed by the reef corals and even in the darkest sunless caves. Here they may be extremely abundant, together with other light-shunning organisms. Some species are quite small, usually one or two inches in diameter. But one of the sclerosponges, *Ceratoporella*, can sometimes reach a diameter of three feet when it lives in caves in shallow reefs. In deeper water this sponge is usually not bigger than ten inches.

Left: A living sclerosponge, Ceratoporella, *at 220 feet on a Jamaican coral reef. Right: A close-up view of the sclerosponge,* Ceratoporella, *underwater at 205 feet on a Jamaican coral reef. Enlarged about 10 times.*

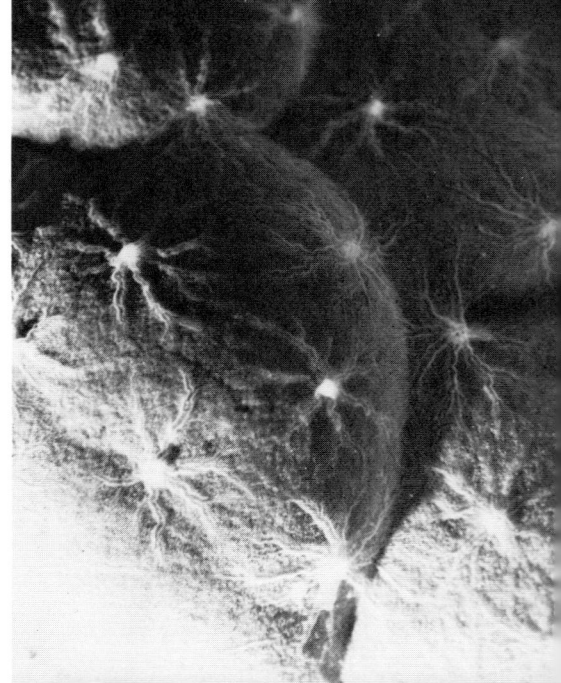

3
The Life of Sponges

Any kind of animal tissue, no matter how it looks to the naked eye, is not really a solid mass. Instead it is made up of millions, even billions, of tiny units called cells. These cells are bounded by membranes and are filled with living matter called protoplasm, plus usually a nucleus or center. The word *protoplasm* is made up of the two Greek words, *protos*, "first," and *plasma*, "form," because protoplasm is believed to be the "first-form" or "matter," the basic element of all life.

Now, we and all other animals are made up of literally billions and billions of these cells. But there are some animals that are so small that they consist of only a single cell. They are able to perform all the life functions using only this single cell. It is believed that these single-celled animals appeared very early in the history of life on this planet. For this reason zoologists call them *protozoa*, a word made up of the Greek *protos*, "first," and *zoön*, "animal." All the animals that consist of many cells are called *Metazoa*, from the Greek word *meta* meaning "after" and *zoön*. They are the "after" or "later-animals," which came after the "first-animals." Thus we see that the Animal Kingdom can be divided into two subkingdoms, the Protozoa and the Metazoa.

It is obvious from their size alone that sponges consist of more than a single cell. Therefore, they must belong to the subkingdom of the Metazoa, and for many years after biologists realized that sponges were really animals, they considered them to be metazoans. But in time scholars began to see that the sponges or Porifera were very peculiar kinds of metazoans, different from all the other metazoans known.

Humans, for example, have a heart, a stomach, lungs, and so on. Each group of cells, united in a special organ, performs a special function like pumping the blood, digesting food, or breathing. In the Porifera the situation is quite different. The cells are not organized into true tissues or organs. There are no such things as sponge hearts, sponge stomachs, or sponge lungs or gills. All the functions performed by the specialized tissues and organs in most Metazoa are performed by individual cells in the Porifera.

This important difference between the other Metazoa and the sponges—namely that sponges do not possess true tissues or organs—is so great that zoologists made a special subkingdom or subdivision for the sponges to which they gave the name *Parazoa*, from the Greek word *para* meaning "beside." Thus, the Porifera are the "beside-animals," belonging neither to the Protozoa, the "first-animals," nor to the true Metazoa, the "later-animals." In the subkingdom Parazoa, there is only one group or phylum, the sponges or Porifera. Because they are so different from the true Metazoa, their basic life functions are also carried out quite differently.

How sponges "eat" and "breathe"

The life functions of sponges, the ways they take in and digest food, absorb oxygen, and eliminate waste materials, are all much simpler than those of most many-celled or metazoan animals. All the functions are based upon a single process: filtering

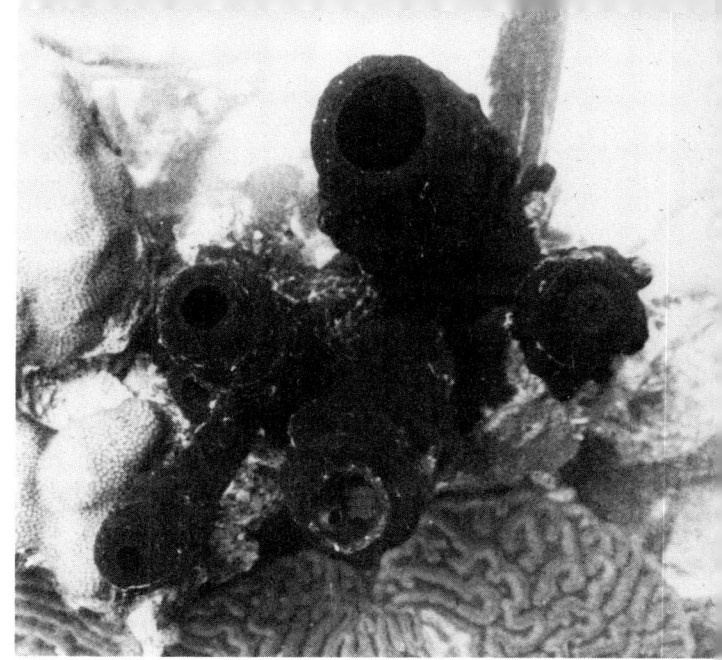

A multitubular specimen of the sponge, Verongia, *showing various stages in the opening and closing of the oscula*

large amounts of water through the pores of the surface and expelling this water through a large opening on the top of the sponge. This opening is called the *osculum,* a Latin word meaning "little mouth."

The water passes through the pores in the outer surface of the sponge and is then forced along through the large number of tunnels and channels. This current is kept flowing by the whiplike waving action of little hairlike structures called *flagella,* from the Latin word *flagellum* meaning "a whip." The flagella are located in a special type of cell called a collar cell, so-called because each cell is surrounded by a tiny collar. As the current sweeps past, the collar cells capture tiny bits of food. Spongiologists have learned only very recently that sponges feed mainly on extremely small particles. Dr. H. M. Reiswig has shown that 80 percent of the filterable material consumed by sponges is so tiny that it cannot be seen with an ordinary microscope. The digestion of food is usually begun in the collar cells and the food is then transferred by moving cells to other cells in the sponge body, where it is put to its final use. No special cells are needed to take in oxygen, since any cell is apparently capable of exchanging oxygen and carbon dioxide, the "waste product" of respiration. The passing current of water also takes up the waste products of digestion, inedible bits of matter and carbon dioxide. This current is pushed through a central cavity and is finally driven out through the large osculum. Thus, a living sponge can be looked upon as nothing more than a large, very efficient filtering machine. Vast amounts of water are filtered in this way every day. It is estimated that eight gallons of water can pass through a medium-sized sponge in twenty-four hours, and in a large wool sponge from the Bahamas as much as two quarts a minute and several hundred gallons a day. Thus, sponges cannot live well in standing or still water; the water around them has to be in continual motion to bring them sufficient food and oxygen. In standing water sponges cease their filtering process and eventually die. Even in well-aerated fish tanks, most sponges do not do very well. However, when left alone in favorable surroundings, some sponges live to a very ripe old age. Although

A huge specimen of the sponge, Verongia gigantea, *photographed at a depth of 120 feet on the reef off Jamaica. This specimen, measuring approximately 20 x 32 inches, may be several hundred years old.*

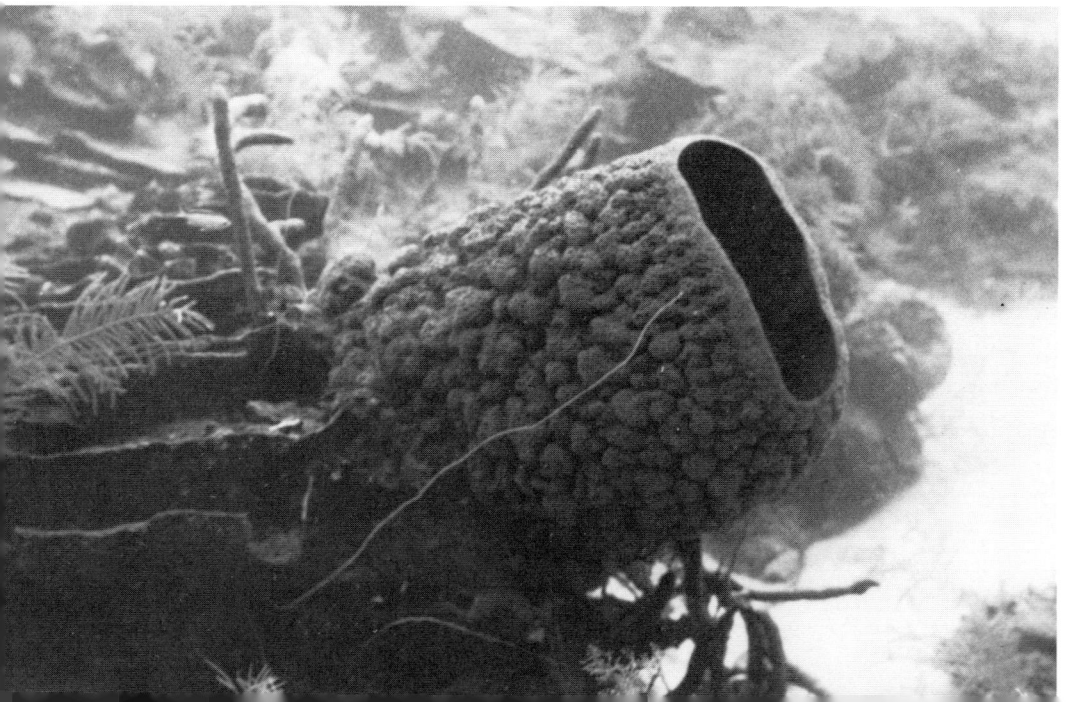

exact facts for most species are not known, some of the larger sponges live to be fifty to several hundred years of age.

THE CELLS OF SPONGES

Cells, as we have seen, are very important in the life of a sponge. They take the place and perform the functions of the important internal tissues and organs of the true Metazoa, such as the lungs, the stomach, and so on. For this reason, it might be a good idea to spend a little time discussing some of these cells. These cells, of course, are all microscopic in size and cannot be seen with the naked eye.

We will first discuss the pore cells which control the size of the pores through which the water is drawn into the sponge. These cells are very properly called *porocytes*, the *-cytes* part of the word coming from the Greek word *kytos*, meaning "a hollow vessel," hence a cell. We will also find this word in the other cell names. The porocytes are shaped like little doughnuts or short bits of macaroni. By contracting or widening the outer ring, they can make the pores larger or smaller and even close them entirely for a short time.

The collar cells, which we read about earlier, are called *choanocytes* from the Greek word *choane* meaning "a funnel," plus the word *cytes*. These important cells do indeed have tiny funnel-like collars of protoplasm just above the main body of the cell. The choanocytes thickly line canals and chambers in the sponge body and, by waving their flagella or little whips, they push the current of water with its food and oxygen past the other cells to be absorbed by them. Then, when the water is laden with waste products, they force the current through to the osculum and out into the surrounding water. As we mentioned earlier, the choanocytes also may begin the digestion of the tiny food plants and animals which they swallow.

The surface of the sponge is protected by a thin layer of hard,

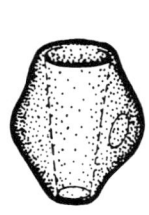 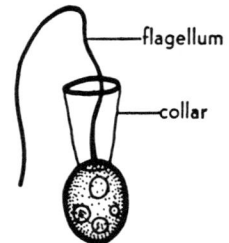 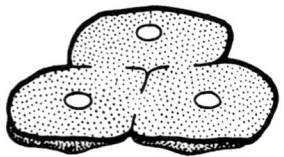

Left: A pore cell or porocyte, much enlarged. Center: A collar cell or choanocyte. Right: A plate cell or pinacocyte

six-sided or hexagonal cells called *pinacocytes,* from the Greek words *pinas-akos* meaning "a plate." Thus, these cells can be called the plate cells. In addition to forming a hard, skinlike protective cover for the sponge, the pinacocytes also have the ability of retracting or pulling in their sides. This reaction occurs very slowly, but it helps reduce the size of the entire sponge when conditions of life are unfavorable or the living sponge is handled roughly.

Another type of cell is of particular importance in the life of the sponge. These cells can wander about within the sponge tissue and bring digested food and oxygen to the other cells of the sponge body and remove their waste products. In this way they can be said to perform the function of the blood system of the higher metazoans like ourselves. Because they are able to wander about and because of their changeable shape, they are called the *amebocytes.* The word *ameba* comes from the Greek word *amoibe* meaning "change," and *ameba* is the name given to a protozoan or one-celled animal which moves about freely by sending out parts of its body called false feet or *pseudopods.* Its body shape has to change all the time while it is in motion; hence, the name.

Now, because the wandering cells in the Porifera look and

move like amebas, they have been given the name *amebocytes*, or ameba-cells. We must mention one type of amebocyte in particular, which will be discussed more in the section on reproduction. It is called an *archaeocyte*, from the Greek word meaning "ancient" or "primitive" and *cytes*. Archaeocytes have a large nucleus and blunt pseudopods. They can form other cells that are needed by the sponge.

There are several other types of sponge cells which we need not discuss here, but the last one we should mention is the muscle cell. Such cells surround the opening of the osculum and can make it larger or smaller or even close it entirely, as conditions demand. They are called *myocytes*, from the Greek word for muscle, *myos*, since they can only expand or contract like a muscle.

How sponges defend themselves

The only defense a sponge has against enemies is passive. It cannot actively strike back and cannot move out of the way. But live sponges for the most part taste and smell so unpleasant that most larger predators have no interest in them as articles of food. Other sponges are poisonous to fish, which quickly learn to avoid them. However, some sponges with horny (spongin) skeletons are eaten by turtles. In other sponges, the presence of sharp and glassy or calcareous spicules within the sponge flesh makes dining on sponges very undesirable and even dangerous. The sponges which nevertheless are eaten by some fish in the coral reef environment are not usually found in the open; they tend to be hidden under coral blocks and in narrow crevices. Some shrimps, sea urchins, and mollusks also prey upon sponges. And then there are a few sponges which can inflict a painful rash on human beings who come into contact with them.

Probably the best defense sponges have is their persistence

in living. Sponges which are torn to bits by storms or other forces do not always die. Many bits simply settle down elsewhere and in time develop into mature sponges.

How sponges reproduce

Sponges have more ways of producing new individuals than most metazoans. Whereas most higher animals reproduce by the union of eggs and sperm (*sexual reproduction*), this is only one of the methods which sponges use.

The formation of a new individual by a way other than the union of an egg and sperm is called, very simply, *asexual reproduction*. This means nonsexual reproduction.

As a sponge grows larger, it may begin to form branches. As these branches develop, each one forms an osculum and other portions of the water circulating system. Actually each branch is capable of becoming a separate sponge, as soon as the osculum is formed and it begins to feed itself. The branches may, however, remain attached to the original sponge. Eventually strong water currents or other forces can break off most of the branches. If the branches can settle down in a favorable spot, they "set up housekeeping" by themselves. In time these new sponges grow large enough to put forth their own branches, and thus new sponges are formed from old. This method of reproduction is called *budding*.

Until the branches are actually broken away, it is hard to tell whether there is really one individual sponge or a group of sponges growing together. We cannot say exactly where one sponge ends and the other begins. For this reason some scientists tend to look upon all larger sponge growths as colonies of several sponges rather than as individuals. Dr. L. H. Hyman, one of the most respected zoologists of this century, has written that the presence of an osculum and water canal system is enough to make an individual sponge. In her view, larger sponges which

A colony of the asconoid, calcareous sponge, Leucosolenia elenor, *from California. Enlarged about 20 times.*

have several oscula may be looked upon as colonies of individuals which are, however, rather hard to separate, one of the important reasons why the Porifera cannot be called real Metazoa and have to be placed in their own subkingdom, the Parazoa.

Living sponges can easily be broken up into little bits by very strong wave action or by attacks by larger animals like fish or crabs. Sponge growers do this deliberately as one way of planting sponges in new beds. The sponge does not die. Each tiny bit of sponge, if it falls upon a favorable spot, can settle down and in time grow into a fully mature new sponge. This method of reproduction, *regeneration*, also shows how primitive the Porifera really are.

The Porifera can also reproduce themselves asexually from units called *gemmules*. (In Latin a small gemstone is called *gemmula*.) A gemmule is formed when many of the special cells

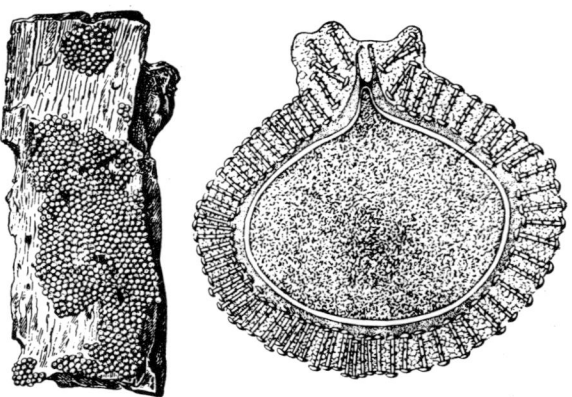

Freshwater sponge gemmules on a board and an enlargement of a gemmule

called archaeocytes filled with food become surrounded by other amebocytes that deposit a hard covering. The gemmules of some marine sponges have a coat of spongin and spicules while others lack the spongin coat.

Freshwater sponges and some marine sponges use the gemmule as a means of survival when living conditions in the water become unfavorable for sponge life. Such conditions occur when the temperature rises or falls to levels at which the sponge cannot live, when the surrounding water becomes highly polluted, or when the food animals and plants begin to disappear. The sponge is capable of forming large numbers of gemmules. When, because of the unfavorable conditions, the sponge begins to die and disintegrate, the gemmule escapes from the sponge and sinks to the bottom. Here it rests until the temperature or the water conditions improve. Then the gemmule opens and the cells begin to develop into a new sponge. Thus the sponge which has "died" is reborn and survives.

All sponges are capable of reproducing sexually. When surrounding conditions are favorable, sperm and eggs, or ova, are formed inside a mature sponge. There are no special organs which form these cells, and both types of cells may be formed in a single individual. In some sponge species, however, the sexes are completely separate, as they are in humans.

In those sponges where both ova and sperm cells are formed, it usually happens that they appear at different times. In this way, the sperm and ova of the same animal will not come together. The sperm of one animal fertilizes the egg of another. This is known as *cross-fertilization* and is the commonest method of fertilization in the Animal Kingdom. The sperm of one sponge is brought together with the egg of another sponge by the same currents of water which also bring oxygen and the tiny animals and plants on which the sponge feeds.

After a sperm has been brought into the body of another sponge by the water currents, it enters a choanocyte or amebocyte. These cells carry the sperm to the egg. We can call these cells "carrier cells." When the carrier cell arrives at the egg, it fuses with the egg and transfers the sperm to it. The fertilized egg then begins to develop into a new sponge. Finally a larva is produced and freed from the parent sponge. The larva can swim around in the water by means of tiny hairs or *cilia*, which it moves around like oars. It attaches itself to a rock, shell, or other substrate and proceeds to develop into an adult sponge.

Compared to other marine invertebrates, sponges have a rather short free-swimming larval stage. The length of this stage

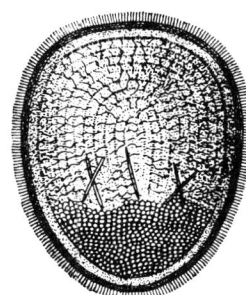
Larva of a freshwater sponge

Release of a cloud of sperm from the demosponge, Verongia archeri

can vary from several hours to about a day, depending on the species of sponge. It is interesting that some demosponges liberate fertilized eggs rather than larvae. In these sponges, the development takes place outside the body of the parent in the seawater.

We said earlier that sponges don't live very well in tanks. Also, sponges always live underwater, so it is hard for us to study them in their natural environment. For these reasons, most of what we know about sponge reproduction comes from sponges that have been collected and that have had tiny pieces of their bodies preserved and fixed on microscope slides for examination. However, in recent years more and more biologists have learned to dive under the seas with SCUBA equipment. In this way they can study sponges and other animals in their natural environments. One such biologist, Dr. Henry Reiswig, was studying the life processes and ecology of some sponges in the West Indies, when suddenly from one sponge there came forth a huge cloud of gray "smoke." He collected some of this material and brought it back to the laboratory. When he examined it under the microscope, he found that it really was made up of sperm from the sponge. This sudden release of clouds of sponge sperm was seen again with other species in the same area. This was the first time that a scientist actually saw sperm coming out into the water currents.

We have seen that sponges have four different ways of reproducing their kind. This richness of methods of reproduction partially explains why sponges appear in such large numbers in many different habitats and why they have existed from very ancient times to the present.

4
Curious Sponges

In a way, all sponges are curious because they differ so much from all other animals that they have a whole subkingdom to themselves. But they are curious in other ways as well.

FRESHWATER SPONGES

People generally think of sponges as being inhabitants only of salt or at best brackish water. But there are many species of sponges which flourish in fresh water, in lakes and ponds in many parts of the world. These are all demosponges and belong to the family Spongillidae. The freshwater sponges are not very large and not particularly showy. Usually, they are merely flattish incrustations on sunken stone, bits of wood, and underwater vegetation. In color, they may be green when they live exposed to sunlight. But this color is not all their own. It is due to the presence of tiny algae which live within the body of the sponge. In the sunlight, the algae, like all plants, flourish and are green in color, but in the shade the algae do not grow very well. The freshwater sponges of India, studied in detail by a spongiologist named Annandale, are rather more showy than those of colder climates. They may assume a variety of interesting shapes. Some produce long, fingerlike branches, others are flat and pillowlike,

A *museum specimen of the freshwater sponge,* Spongilla, *growing around twigs*

and some, like the species of *Verongia* in coral beds in the sea, are covered with ridges. They are also a little brighter in color, but never as handsome as the sponges in the coral seas.

Freshwater sponges rarely grow very large. They are exposed to more extreme and harsh conditions than marine sponges, whose environment is much more stable. In the colder lands, freshwater sponges cannot survive the bitter cold of winter, and in the tropical countries they perish in the suffocating heat of summer. When they are about to give up their normal shape—we cannot say that they actually die—they form gemmules as we described in an earlier chapter. Though the rest of the sponge body then perishes either in the cold or the heat, the gemmules survive and begin to grow again into mature sponges when the warm weather returns in the temperate zones, and cooler weather in the tropics.

Freshwater sponges, like most species of sponges, are of no commercial value. Other creatures, however, find them quite useful. Algae, as we have seen, and some tiny insects use them for shelter. Certain insects lay their eggs in the freshwater sponges and there the young hatch out. Some snails, crabs, and fish find freshwater sponges very much to their taste. Thus, like all organisms in nature, the Spongillidae too find their own useful place, even if we humans do not make direct use of them.

THE BORING SPONGE—*Cliona*

It is hard to imagine that a creature like a sponge—soft, unable to move from place to place, and considered for centuries to be plantlike in nature—can actually bore into stony corals, shells, and calcareous rocks. On the beach one frequently finds shells or fragments of shells completely riddled with a large number of channels and galleries like minute coal or silver mines. This is the work of the excavating sponges of the genus *Cliona*. *Cliona* is known to excavate only calcareous materials and it is found in many different parts of the world. It is more abundant in warmer seas where there is a great deal of calcareous material—shells, coral, and other kinds—it can bore into.

Cliona doesn't look much like a sponge. While it is excavating galleries in a shell, for example, one can see only small rounded bits of flesh sticking out where holes have broken through to

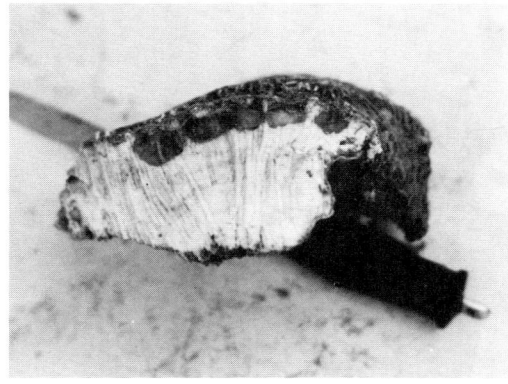

Excavations produced by Cliona laticavicola *in a massive coral head*

the surface. When it is mature, the sponge may form a thin layer over the shell it has been excavating.

Here is how *Cliona* operates: the larva of a *Cliona* settles on a snail, clam, or oyster shell. Then it begins to dig out tiny chips of calcium, in a manner which is not yet completely understood, until it finally succeeds in excavating long, twisty channels and galleries. In time parts of the sponge appear on the surface of the object it is excavating. One common species called *Cliona celata* (the second name comes from the Latin verb *celare* meaning "to hasten") begins to look like a lot of yellow pimples sticking out of the calcium surface. Because of this color, it is often called the sulphur sponge. Later these pimples begin spreading over the entire surface, and finally the whole surface is covered with a thin, smooth layer of *Cliona* sponge. This process of covering over the surface is called "encrusting." In time all the shelly material disintegrates and only the tough sponge matter is left.

Cliona's habit of excavating galleries in calcareous material is very useful to nature. It plays an important role in removing rubble from the ocean floor and recycling the calcium found in it.

In the course of time millions, even billions, of shell-bearing mollusks and other shelled organisms must die and leave their hard, calcareous parts behind. Corals also die and add their chalky skeletons to coral reef formations. These reefs serve as a framework on which many animals can find homes, but they also contain vast quantities of calcium that were taken from the sea while the corals were building their skeletons.

By boring holes and galleries in shells and corals, *Cliona* weakens them and causes them to fall apart. In this way the dead litter is eventually broken down and removed. The formerly useless calcium can be recycled by new generations of living mollusks and coral animals to build new shells and coral reefs, and nature's cycle can continue. Although there are other marine animals that break up calcareous material, the excavating habit of *Cliona* is an extremely important one.

However, as so often happens in nature, *Cliona* can cause some harm as well. The sponge larvae sometimes settle on the shells of living clams and oysters and begin to construct their tunnels and galleries there. As the sponge begins to penetrate the inside surface of the shell, the poor clam or oyster tries to protect its soft body by putting down new layers of shell material. But to no avail. The sponge continues its excavating and eventually the weakened bivalve can no longer protect itself and perishes.

Cliona can also cause harm to growing coral reefs. Delicate coral heads growing on a reef have their bases eroded away by the boring sponge at the points where they are attached to the solid reef. Eventually they break off and fall down steep slopes

Left: An oyster shell pierced by Cliona, *the excavating sponge. Right: The excavating sponge,* Cliona delitrix, *attacking a head of star coral. Notice the large tubes of a* Verongia *sponge in the foreground.*

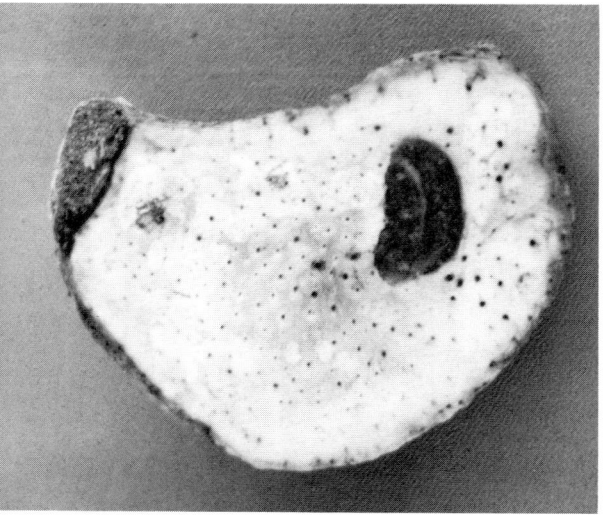
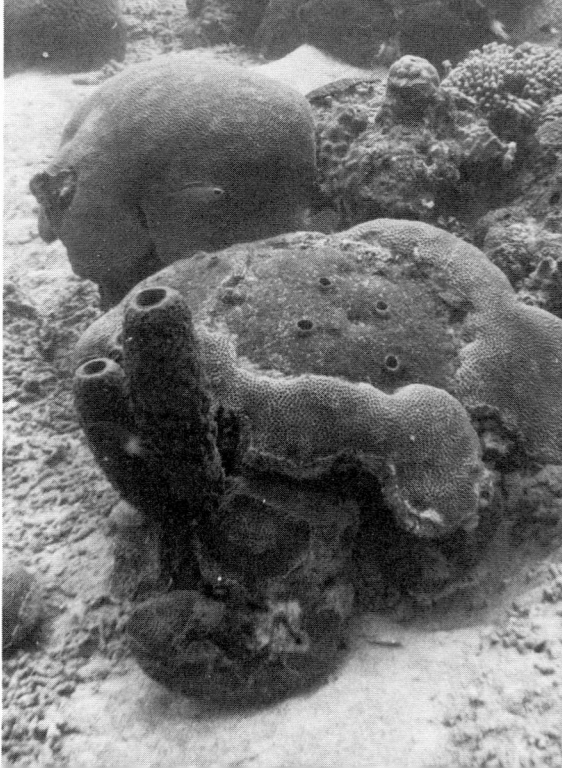

The "siphon-net" sponge attacking some star coral. The photograph was taken at 75 feet underwater on a coral reef off Jamaica.

to the ocean depths where, away from all sunlight, the little coral animals die. In this way the boring sponge controls to a certain degree the shape and growth of coral reefs. Few reef dwellers exercise so much power.

The damage which *Cliona* does to shells and corals can also be done to seawalls built of calcareous rock or coral boulders. The cities whose harbors are protected by such walls, are put to continuous expense to keep the walls in repair. *Cliona lampa*, a brilliant red or vermilion-colored excavating sponge, covers huge areas (sometimes several hundred square feet) of chalky seawalls and calcareous rock in Bermuda. Tourists there are surprised to see part of the seawall looking red or pink, but they little suspect that a certain amount of damage is being done to the wall.

While we are talking about how destructive excavating sponges can be, let us mention that a spongiologist at the Smithsonian Institution, Dr. Klaus Ruetzler, recently reported that some excavating sponges called *Siphonodictyon* (the name means "siphon-net" in Greek) make very, very large, rounded excavations and do a lot of damage to the skeletons of the corals they excavate. Most species of *Cliona* form only small or tiny galleries that look like honeycombs. The "siphon-net" sponge is much more destructive. Although *Siphonodictyon* behaves like the *Cliona* excavating sponges, it belongs to a separate family called the Adociidae.

Thus *Cliona* and other excavating sponges can cause some damage to the things mankind is interested in. But when we draw up the account of *Cliona*, we are forced to admit that on balance it causes more good than harm. Clams, oysters, and living coral have many more serious enemies than *Cliona* and they kill many more individuals; the calcium walls of harbors are attacked also by boring clams and other creatures, and the amount of seawall and coral destroyed by *Cliona* is not very large. The good it does by helping disintegrate dead shells and coral blocks on the sea bottom and recycling the calcium is far more important to us and to the sea creatures which are useful to us. If *Cliona* were to vanish altogether, we would be very much worse off.

Stinging sponges

Some kinds of sponges have a direct effect upon the skin of unwary SCUBA or skin divers who swim in a coral reef. For convenience we can call these the "stinging" sponges. We must remember that they don't actually sting like a bee or even like a jellyfish. They produce substances that are irritating or stinging to the skin. Two of these sponges, called *Neofibularia nolitangere* (the second name means "do not touch" in Latin), a brown or red-brown sponge, and *Tedania ignis* (the second name means "fire" or "flame" in Latin), which is orange-red, can make the skin very red and swollen. When human skin comes in contact with this harmless-looking sponge layer, it feels a sharp, burning or prickling sensation. Even when the owner of the skin veers off immediately, the pain does not diminish. Days later the affected area is still red and swollen and very painful. It seems that some people are more sensitive to these sponges than other people, just as some people are more allergic to pollen or roses. The skin rash can be treated by antihistamines, like allergies. Another interesting fact about these two sponges is that, although they can cause humans great pain, coral reef fish have been known to eat them and not suffer any consequences.

This ability of the sponge to cause so much pain is probably an adaptation on the part of the sponge to protect itself from some enemies. Chances are that the diver who suffers from a sting will be more careful in the future. He'll keep a sharp watch out for the innocent-looking little stinging sponge and keep out of its way. Many animals will also react in this way.

Poison sponges

There are some sponges which are very poisonous to the fish which eat sponges. Some fish dine regularly on sponge meat. But when these fish are forcibly fed bits of poisonous sponges, they soon die.

Things are made especially hard for the sponge-eating fish. In Veracruz, a harbor city in Mexico, some spongiologists noticed that by far most of the species of poisonous sponges live openly exposed to fish bites, while the edible sponges are usually well-hidden under coral heads and in narrow crevices. The fish in Veracruz have to learn very quickly that an easy sponge meal can be deadly. They may have to hunt harder to find good food. Although we know that some sponges are poisonous to the fish that eat them, it is surprising to see that a few species of tropical fish can actually eat poisonous sponges in their natural environment and survive. It is interesting to note here too that a sponge commonly known as "the stinker" (*Ircinia strobilina*), because we humans find its odor so awful, is actually eaten by some fish.

There is another spectacular example of a poisonous or toxic sponge. A biologist named P. G. Bryan, working in Guam, began to study a grayish encrusting sponge which seems to be a new and unnamed species of the genus *Terpios*. He found that this sponge could grow over dead coral and shells and also over live corals. When he took the sponge and some living coral into the laboratory and placed them side by side in a tank with running seawater, some of the coral tissue was killed by the sponge within a very short time, and the rest of the coral drew its tentacles tightly into itself and closed down its activity as much as possible. Dr. Bryan thought that the sponge might be getting some nutrition from the coral tissue because the sponge grew faster over living coral than over dead coral skeletons. We know of course that the sponge can't chew off pieces of coral tissue and use them as food. However, the sponge is indeed toxic to the coral and may pass pieces of coral tissue it has killed through its own filtering system. How this actually works no one knows yet, but the very idea of a "killer sponge" is both scary and fascinating.

THE STRUGGLE FOR LIVING SPACE

Many species of sponges fix themselves to a solid base and grow upward. But there are also many others, the encrusting sponges, that grow like a layer of moss, not increasing in height but spreading out over a larger and larger area like a blanket. Animals that are able to move, even slowly, can keep away from the spreading blanket and stay alive. But others that are sessile or fixed to one spot, like oysters, can only wait while the sponge slowly covers them and finally suffocates them by shutting off their supply of water-borne food and oxygen.

These encrusting sponges are able to overcome, in their quiet but persistent way, many competitors for the restricted free space. But chances are that most of the animals and plants which lose out to such sponges are not of great direct importance to mankind.

In spite of this intense struggle for living space in which one of the two competitors usually dies, there are some cases where both the encrusting animal (the encruster) and the encrusted animal (the encrustee) survive. Biologists studying the ecology of Jamaican coral reefs observed little tubes, bright lemon-yellow in color, sticking up like chimneys out of the surface of corals. They realized that these were the oscula of a sponge and named it the little chimney pot sponge—*Siphonodictyon brevitubulatum* in zoological language. Sometimes a bright red-orange encrusting sponge grows over the coral and right up to the tubes of the little chimney pot, which, however, keeps right on growing. Picture in your mind the colorful image of the coral, which may be brownish yellow or even green, with the lemon-yellow tubes of the chimney pot surrounded by the red-orange encrusting sponge. Here is an instance where the struggle

The encrusting sponge, Mycale laevis, *and a reef coral,* Montastrea annularis. *Here the sponge has surrounded the coral and actually divided the living coral surface into several separate units.*

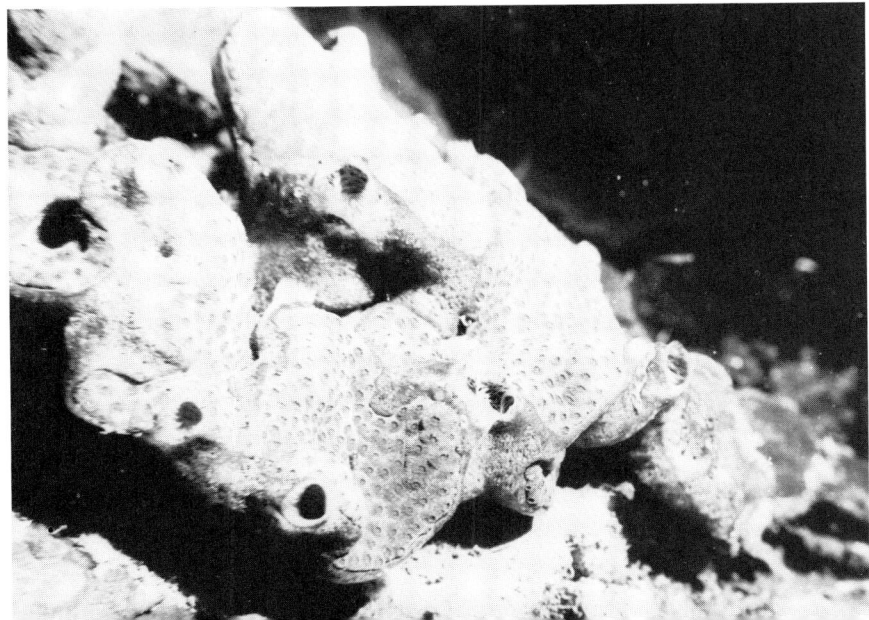

for space does not result in the death of one of the competitors; both sponges survive.

There is another encrusting sponge called *Mycale laevis* which lives on coral formations in the Caribbean Sea. Drs. Hartman and Goreau, whom we mentioned in connection with the discovery of the Sclerospongiae, are also responsible for this interesting observation. *Mycale laevis* encrusts the lower surface of some reef corals that normally are flattened or platelike. When the sponge reaches the edge of the coral, the rim of the coral turns upward, forming little arches. The oscula of the sponge open at the top of these arches, and apparently the edge of the coral grows upward because of the strong water currents coming from these oscula. This association has advantages for both the coral and the sponge. The encrusting sponge has all to itself the under surface of the coral that is always growing bigger, and the coral is protected from the invasion of *Cliona*, since its bottom surface is completely covered by *Mycale laevis*.

SPONGES AND CRABS

There is an interesting relationship between some members of a family of demosponges, the Suberitidae, and hermit crabs. The larva of certain suberitid sponges may attach itself to an empty snail shell in which a hermit crab has made its home. Here the sponge develops and continues to grow until both the

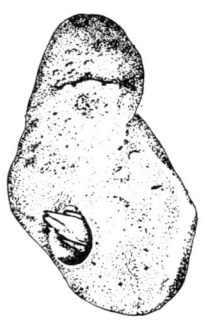

A suberitid sponge inhabited by a hermit crab

shell and the crab are entirely enclosed by the sponge, except for the shell opening through which the crab can poke its head and legs in order to obtain food and oxygen and move around. Eventually all the shell matter dissolves and the hermit crab finds itself living inside a spiral cavity with walls of smooth, firm sponge tissue. It must be a curious sight to see the crab peeping out from a lump of sponge and carrying it around wherever it goes. This relationship is beneficial to both the sponge and the crab: the crab finds protection from its fish enemies inside the firm walls of the bad-tasting sponge; and the sponge benefits because it is moved from place to place and can filter water in different localities. In fact, it seems that some suberitid sponges are only found on snail shells inhabited by hermit crabs.

Another crab, *Dromia vulgaris*, is called the sponge crab because of the peculiar use it makes of sponges. The sponge crab has a broad, hairy shell about three inches wide. The last pair of the crab's legs are bent upward so that they can tear off bits of living sponge matter and hold them against its hairy back until they become fixed there. In a short time these bits of sponge begin to grow and eventually cover the entire back of the crab, fitting closely over it like a cap. Other marine organisms also become attached to the sponge mass so that in time, as one writer says, "a crab bedecked in this manner looks like an animated submarine garden on a small scale." No hungry crab-eating fish or other predator would ever suspect that under this mass of spicule-filled, evil-smelling sponge mass covered by other exotic organisms there rests a quantity of toothsome crab flesh. The well-hidden crab is thus protected from his enemies. When the crab molts and changes its old shell for a larger one, it has to go through the same sponge-planting process again. The mutual benefits which both the crab and the sponge receive here are very similar to the benefits of the hermit crab and his sponge shelter.

Sponge condominiums

There are literally billions of small creatures in the sea which need places to shelter themselves, lay their eggs, and brood their young. And few places are so advantageous to them as a large sponge. Huge numbers of small fishes, serpent stars, mollusks, and other marine animals use such sponges as condominiums or apartment houses. One zoologist once counted, in addition to large numbers of other creatures, exactly 16,352 shrimp living in a single large loggerhead sponge in the Tortugas. To see how such creatures can live inside a sponge, it is necessary only to collect a medium-sized one and bang it lightly a few times against a board. What a lot of excited scurrying, jumping, running, hiding, wriggling, and scooting about takes place as the evicted tenants frantically look for other shelters. The nice thing is that the sponge tenants seem to do their sponge hosts no harm at all.

There are also some kinds of fish, gobies and blennies, which live inside sponges. They are usually small, long, and slender and get around easily inside the tunnels and channels of the sponge. Here they may spend their whole lives, catching their food, laying their eggs, and hatching and guarding their young. Barnacles and sea anemones which do not live inside the sponges

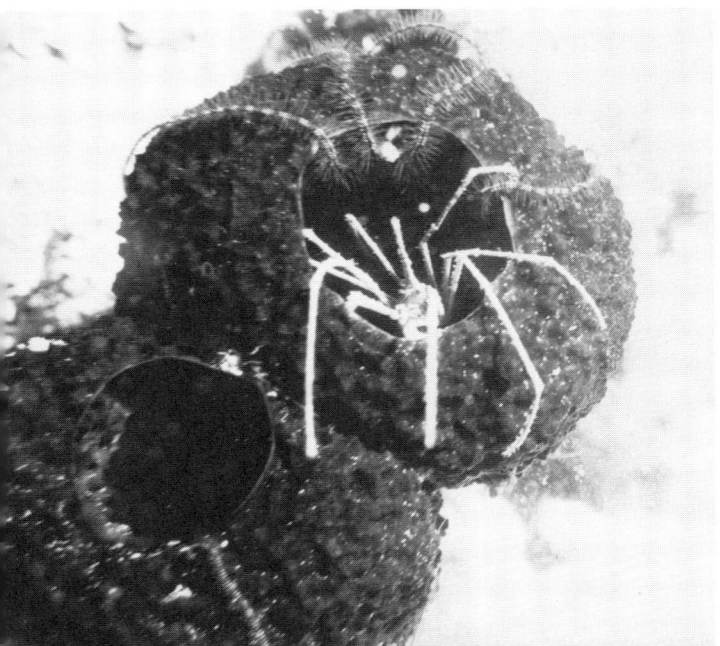

Some sponge tenants: an arrow crab and several brittle stars in the osculum of a Mycale *sponge*

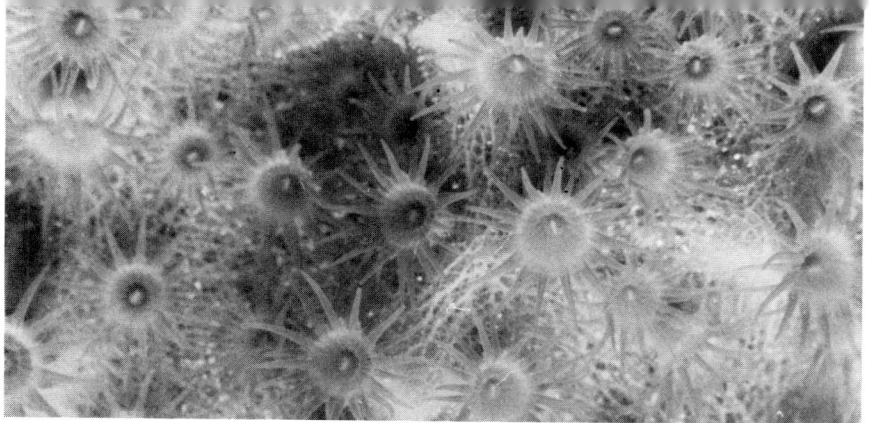

Zoanthids on the sponge, Gelliodes. *Enlarged 10 times.*

attach themselves to the hard, outer layer. These also do not seem to bother the sponge to which they are permanently fixed.

SPONGES AND OTHER ANIMALS

There are two other sea creatures which make interesting uses of sponges. One is a tiny creature which looks very much like a sea anemone or coral, to which it is related. It is called *zoanthid* meaning "animal flower" in Greek. Some, which are so small that they look like tiny white dots, live fixed to the body of an excavating sponge called *Cliona delitrix.* As the sponge digs away in the coral on which it lives, it frees tiny bits of chalky material. The zoanthid captures these pieces and, together with spicules formed by the sponge host, it constructs a tiny protective case or coat for itself. Thus, the spicules which the sponge forms for its own skeleton also serve another sea creature for a different purpose.

Finally, there is a kind of sea slug which lives on a brilliant red sponge. It is one of the predators which likes sponge flesh and eats bits of the sponge when it is hungry. In order to protect itself from its own enemies, the slug has the same color as the sponge so that a slug predator hunting for food will pass it by, taking the slug for part of the unappetizing sponge. Thus, the sponge not only provides the slug with food, but also with protective coloration.

The hairy sponge, Tethya crypta, *suffering continuing predation by the sea urchin,* Eucidaris *(foreground)*

Food sponges

Most predatory animals prefer not to eat sponges. They are repelled by their generally unpleasant odor and taste or are scared off by the sharp, needlelike spicules. And some sponges, as we have seen, are actually poisonous. However, there are certain types of crabs, fish, shrimp, sea slugs, sea urchins, and snails which relish sponge meat. It has been found that some fish have a diet actually made up of 85 percent sponge.

Some crabs, fish, and other animals can attack a sponge directly, while smaller animals have to wait until a sponge is injured. Only when the hard outer layer composed of pinacocytes is broken, can they get to the soft sponge material inside.

Sponges have rarely been eaten by human beings. However, the small creatures which do nourish themselves on a sponge diet are eaten by larger creatures, and these eventually by still larger ones like fish, which we consume. In the long run, unpalatable sponge tissue does serve as nourishment for us humans.

FALSE SPONGES

Sometimes in a drug store one can buy a long, narrow, sponge-like object which is called an Australian sponge or a Japanese sponge. Its real name is *luffa* (sometimes written *loofah* or *loofa*) and it is not a sponge at all. It is the inside of a long, cucumberlike gourd which grows on a vine in Florida and in other warm and hot regions of the world. People use it like a sponge and rub it vigorously over the skin while bathing. It is not surprising that they take it to be a kind of sponge. Its tough, interwoven, stringlike fibers look very much like some kind of sponge material and the large tunnels running lengthwise through the center look like the body cavities of sponges. Nevertheless it is only a vegetable and does not belong to the Porifera.

All in all, even harmful sponges are only marginally harmful to human beings. The boring sponge in the long run is more beneficial than harmful; the suffocating sponges are more dangerous to sessile organisms than to us; the poison sponges deprive us of only a few careless fish; and it is as easy for a diver to avoid being hurt by a stinging sponge as it is for a hiker to avoid nettles and poison ivy in the woods.

There is something that we must realize not only in connection with the Porifera but with all our fellow creatures, big and small, on this planet. The stinging sponge and the boring sponge *Cliona* and the others which we have mentioned in this chapter are not aware of the "harm" or the "good" they do. They merely follow their instincts which have evolved over eons of change. It is we humans who determine whether what we do is "good" or "bad." In the story of nature, there really are no heroes and no villains.

5
Sponges and Man

Basically, sponges are useful to mankind because nothing else in nature can sop up so much liquid and release it so quickly by simply being squeezed. Thus they are much used for washing and similar tasks. But there are a number of other uses for sponges which might not be so well known. Before we take up these less well known uses of sponges, let us see how commercial sponges are harvested and prepared for sale.

Sponge fishing

All commercial sponges, whether fished in Florida, Mexico, the West Indies, Japan, or the Mediterranean Sea, belong to the family Spongidae of the class Demospongiae. They all have a leuconoid body made of flexible spongin and do not normally have spicules. They vary mainly in the softness of their spongin, the softest and most expensive being the kind called silk sponges from the Mediterranean.

The commercial sponges usually have both scientific and common names, since they are the sponges mankind is most familiar with. The common names for these sponges, used by sponge fishermen, are quite descriptive. Some examples, in addition to the silk sponge, are the wool sponge, the elephant's ear

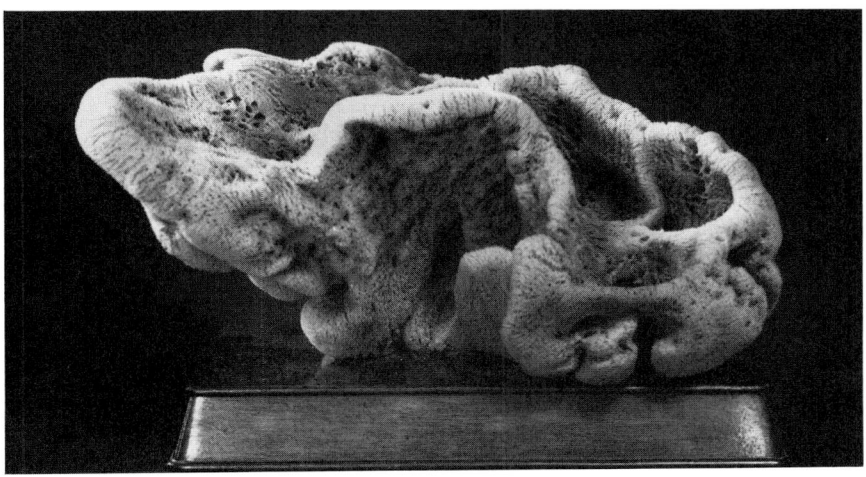

The Syrian silk sponge, one of the softest and finest of the commercial sponges

sponge, the velvet sponge, and the honeycomb sponge.

Fishing for sponges is an ancient and simple process. Basically, it consists of finding the sponges and tearing them loose from their foundation. When sponges grow in shallow water, the simplest way is to wade in and pick them off the bottom. Sponges living just a bit deeper can be hooked. This is done by two men in a rowboat, one who rows and one who hooks. The latter holds a long pole with a three-pronged hook at the end in one hand and in the other a water glass. The water glass is simply a bucket in which the opaque bottom has been replaced by a piece of strong plate glass. When the bottom of the water glass is held an inch or so below the surface, the ripples on the water are smoothed out and the sponger has a clear view of the sea bottom. When he sees a sponge, he uses the hook at the end of the pole to tear it loose and haul it aboard. These men, who are called "hookers" in the industry, can collect sponges in depths of about 30 feet. When sponges live in even deeper water, diving becomes necessary.

From very early times the Greeks of the Aegean Sea found that sponges in their part of the world grew in fairly deep water, and they used a method called "free diving" to obtain them. This just meant that the diver would plunge into the water, holding his breath for as long as possible while he searched for the sponges. This type of sponge fishing is still used today, but it is quite a dangerous occupation and requires great powers of endurance.

Diving with a helmet is another way to collect the sponges in deep water and is still widely used. The diver puts on a heavy suit with a big head cover that looks a little like a space helmet. In such a suit, to which air has to be pumped to keep the diver alive, he can walk along the bottom as deep as 125 feet, though usually he works only at 70 to 90 feet. The diver tears the sponges loose with a short-handled hook and deposits them in a large net bag. When the bag is full, he signals to the people on the sponge boat who haul in the catch. This is a difficult way of har-

A sponge diver

vesting sponges. The equipment is quite expensive and the diver has to be very careful. If he is pulled up too rapidly from the great pressure at the lower depths to the reduced pressure at the surface, he can suffer from a terrible condition called the "bends," which can cripple him for life or kill him outright.

Once the sponges are aboard, however, they have to be cleaned before they can be used. The sponges are placed in piles on the deck of the sponge boat and are covered with wet cloths. Here they stay until they die and begin to decay. The sponges are then put into washtubs full of seawater and are pounded and stomped on by crew members wearing rubber boots. The material that comes out—the half-decayed sponge flesh and dead and dying creatures living in the sponge—is called "gurry." It is not the sweetest smelling mixture in the world. When the sponges are sufficiently cleaned of the gurry, they are strung up like huge beads on a heavy cord and hung on the masts to dry. Sponge boats returning from successful sponging voyages are literally covered with such sponge "necklaces." On shore the sponges are trimmed and rough bits of coral and stone cut away. Sponges which are to be used for rough work by painters and car washers are now ready for sale, but bath sponges and those used for more delicate purposes are bathed in chemicals and bleaches to make them softer and their color more pleasing.

Sponges can also be harvested by dredging. But this is a very wasteful method since it tears up everything from the sea bottom. As a result many other creatures like small fish, crabs, oysters, and so on are destroyed. This method has long been forbidden by the United States government. The government also protects the sponge industry by forbidding the capture of any sponge less than five inches in diameter when alive. Thus, the small sponges are preserved and allowed to grow to a bigger size.

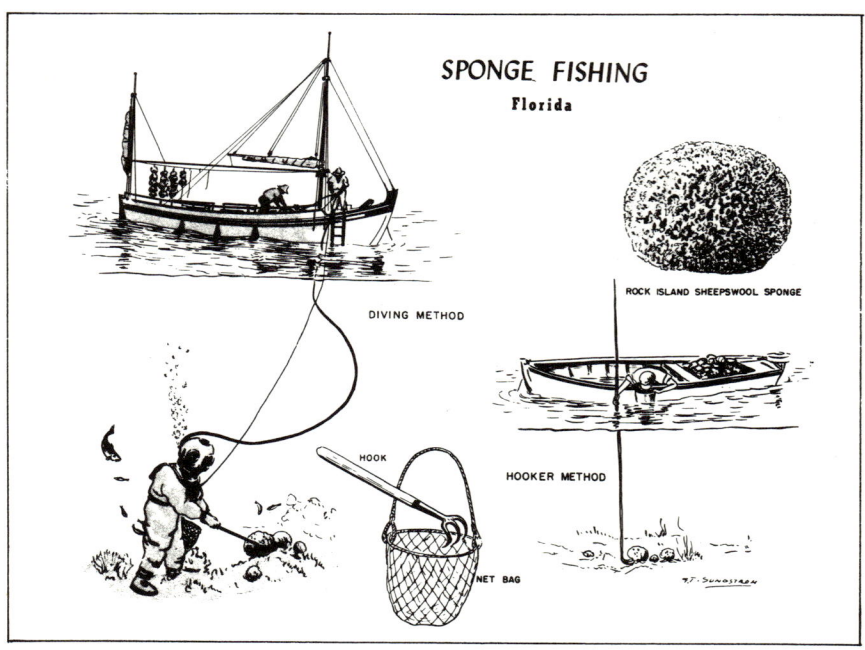

Sponge fishing in Florida

Sponges have also been harpooned, like whales. But this method tends to tear the sponges and make them less marketable. However, conservationists see an advantage in harpooning, because torn bits of sponge are left behind and these, in time, can grow into new mature sponges.

Finally, sponges can be collected by SCUBA diving, a method that does not seem to be used much in the sponge industry. The reason is probably that by the time the SCUBA gear was invented and perfected, sponge fishing had become a much less profitable industry and not attractive to the divers.

As we shall see, natural sponges are very important in some industries. This situation will probably remain so for many years to come. Nevertheless, sponge fishing, except in a few localities, is not a flourishing industry today. The city of Tarpon Springs in Florida, which used to be one of the most important centers of the sponge industry in the world, in a peak month in 1973 produced less than 750 pounds of sponges. At present a large part of its income is due to tourists who come to see how the sponge industry used to be conducted. In other parts of Florida also, the sponge industry shows signs of declining. From January to June 1972, for example, a little more than $95,000 worth of sponges was landed, a small enough sum in itself. But even this small sum sank by about one-half to little more than $50,000 in the same period in the following year.

One of the problems is that the sons of the old sponge divers do not want to continue the work of their fathers. They find it too dangerous, too laborious, too risky, and not as profitable as many other types of work. The result will probably be that sponging will die out in the United States and be concentrated in regions where young people do not mind working at it.

Sponge disease

In addition to all their other problems, the sponge fishermen have to contend with epidemics that affect sponges. After many years of good profit came the sad years between 1943 and 1949 when the deadly "red tide" attacked Florida's beaches and dealt the sponge industry a serious blow.

The "red tide" is caused by a microscopic protozoan called *Gymnodinium*. It is always present in the water, and in small numbers *Gymnodinium* can be disregarded. Sometimes, for largely unknown reasons, the population increases greatly— "blooms," as ecologists say. Then the tiny bit of nerve poison which each individual *Gymnodinium* gives off reaches a tremendous concentration. As a result, all creatures in the affected water are killed. The usually clear, beautiful, green and blue

Idle sponge boats in Tarpon Springs, Florida

waters of Florida turn reddish, muddy, and cloudy, and the beaches are littered with dead and dying fish, mollusks, worms, and everything else that is usually found near shore. In windy weather, tiny droplets of the "red tide" waters are blown ashore and irritate the noses and throats of people, causing them to sneeze and cough.

The red tides which came between 1943 and 1949 were particularly widespread and disastrous for the sponge industry. The sponges sickened and died. Some sponges lost their hard protective layer, which is formed, as we know, by the cells called pinacocytes. In this condition, the sponges turned foul and gave off an evil odor and soon died. In others the outer layer was left on, but at the touch of a sponge fisherman's hook, the sponge fell apart. Sponge fishing almost came to a complete halt. People began using the synthetic sponges. Since these were cheaper, the superior natural sponge lost much of its market, which it never regained.

The sponge population, as that of most other creatures, eventually recovered from the attacks of the red tide. At present, the sponge beds are again fully populated with large, healthy sponges. But they are not being harvested, which in itself is a hazard. An unharvested sponge bed tends to become overpopulated and overcrowded. In such conditions the sponges do not reach their largest and most profitable size. Thus, a vicious circle is formed. The sponges are not harvested because it is not profitable to do so. As a result the unharvested sponges lose their desirable shape and size, and it becomes even less profitable to harvest them. Then the sponge beds become even more crowded and the quality of the sponges becomes even worse, and so it goes. The chances of a new and profitable sponge industry appearing in this country become more and more unlikely.

In spite of its many problems, the sponge industry will never entirely die out. Even if it disappears from one part of the world, it will continue elsewhere because there will always be a need, even if only a small one, for natural sponges.

Painters, window washers, tile men, ceramic and pottery makers, lithographers, jewelry makers, decorators, and many others prefer natural sponges. For this there are good reasons. Natural sponges last longer even under heavy wear. They do not tear or "peel away" as easily as the manufactured sponges. Because they have many more chambers and tunnels, they can soak up much more water without dripping and they can, for the same reason, also be washed clean much more quickly and easily. They are in general softer and in the long run can turn out to be more economical than the other kind.

Sponges in ancient times

Many centuries ago the ancient Greeks and Romans used sponges for bathing and cleaning chairs and tables very much as we do today. They also used them as a lining or padding for their war helmets and armor, so that these things would be easier to wear in battle. The Romans made paintbrushes of sponges and tied them to long sticks to be used as floor mops. The poorer people and soldiers also had to use sponges for drinking wine. Each soldier dipped a sponge into a large common bowl of wine. Then, holding the sponge up in the air, he squeezed the sponge and let the wine run into his mouth. This may not have been a very tidy way of drinking wine and probably left the hands sticky. But if there were no handy wine cups around, it was probably the best way. We read in the Bible that a Roman soldier held a sponge soaked in vinegar to Jesus on the cross. This vinegar was actually something which the Romans called *posca*. It was made by mixing equal quantities of vinegar and water and was very popular among the poorer classes in Rome who could not afford to buy real wine. It seems that the Roman soldier offered the drink in the only way he knew, with a sponge.

SPONGES IN MEDICINE

Surgeons use many sponges in their operations. In early times surgeons used real sponges. These sponges had to be the finest and softest known and so generally came from the Mediterranean Sea. In more modern times, other, artificial materials came into use as "sponges." Most often these are folded, sterile gauze pads. Although the old use of real sponges is still mentioned in books on the history of medicine and in medical dictionaries, surgeons who have been trained in recent years have never even seen a real sponge in the operating room.

Sponges were also used as folk medicines long ago. In the Middle Ages people burned sponges and used the ashes as medicines. More recently bits of cut-up sponge were applied to wounds to make them heal more rapidly. In the Midwest and in mountain areas of this country many people often suffered from goiter. This was often caused by a lack of iodine in the diet. People who lived in these regions many years ago were not able to eat enough seafoods like marine fish, clams, shrimps, or lobsters, which are very rich in iodine. Nowadays people know that they require iodine in their diet and they use salt to which iodine has been added. Formerly, people in regions very far from the sea might be given finely ground bits of sponge to drink. The sponge pieces must have had enough iodine remaining to help cure the goiter. And it probably was the iodine in the sponges which also helped cure wounds and infections.

Recently, researchers have been much excited by an antibiotic material which they have found in sponges. They have named it *ectyonin,* and in time it might become as important in curing people as penicillin, sulphanilamide, or aureomycin. If it does, it might help revive the sponge fishing industry throughout the world.

Sponges and cosmetics

In stores specializing in the sale of cosmetics, one can buy small bits of sponge which are used to apply rouge and powder to the face and to remove it at night. Such sponges are also from the Mediterranean, though some of the softer Florida and West Indian sponges can be used. Actors may even use such sponges to put on and remove their stage or TV makeup. Actors sometimes feel sharp stings as they use the sponges. These stings can be caused by tiny bits of glassy spicules. Although commercial sponges of the family Spongidae do not produce spicules, some spicules or bits of spicules from other sponges, together with bits of shell and coral, can become lodged in the sponge fiber and have to be removed. Sometimes this is not done thoroughly enough.

Sponges as wedding presents

The fact that other animals like to live inside sponges, and sometimes pass their entire lives there, led to a pretty custom among the Japanese. A certain type of shrimp lives in pairs, one male and one female, inside the lovely Venus's flower basket sponge. The shrimps enter while they are still small. When they grow big, they find that they can no longer leave the sponge because the osculum of the sponge is covered by a strong, sievelike plate. Thus, they spend the rest of their lives inside their beautiful glassy home. Apparently they thrive in such an environment. The scientific name of these shrimps is *Spongicola* which, fittingly enough, means "sponge-dweller" in Latin. In the past, the Japanese collected sponges containing the shrimp couple and gave them as wedding presents to newly married human couples. They felt such a gift symbolized a marriage which would last until death.

Fossil sponges

Most people, when they think of fossils, think first of all of the huge skeletons of dinosaurs and mastodons which are set up in museums. But invertebrates, animals without backbones, also leave fossils. These are not bones, but shells and other hard parts of their bodies. The bodies of most sponges are made of rather weak material which decays and disappears soon after the sponge dies. Thus, whole sponges are frequently not preserved as fossils. However, as one can expect, the hard spicules of the sponge do remain, and the earliest sponges are known only from fossilized spicules.

Fossilized spicules are often the chief component of the mineral called flint. In some places in the world, in prehistoric time, sponges were so abundant that the spicules they left behind were numerous enough to form thick deposits of flint. At times even complete sponge skeletons are found in the beds of this useful material, useful because flint was the chief material of which primitive man made his stone implements. The American Indians made their arrowheads of flint. It is curious to imagine that the Indians could have killed a buffalo or deer, or

A reconstruction of a scene from very early seas during the Cambrian period. Notice the tubular sponges and the strange-looking crustaceans.

even a human enemy, with nothing more than a mass of fossilized sponge spicules.

Pieces of flint were also used to make fires because they gave off a strong spark when they were struck. Among the earliest rifles and pistols invented were some called flintlocks, because a bit of flint in the hammer struck against a bit of metal. The spark lit the primer and eventually set off the powder which fired the ball. Thus, one can say that not only have some living sponges been useful to man, but some fossilized sponges from ancient seas as well.

6
Working with Sponges

A SPONGE COLLECTION

Many people who are interested in ocean and seashore life like to make collections of starfish, sand dollars, sea shells, crabs and shrimps. Very few collect sponges. But there is no reason why a sponge collection cannot also form a very interesting display and absorb the attention of a nature lover. Now that SCUBA diving is becoming such a popular hobby and as people learn more about sponges, collections of the Porifera may become more widespread.

In very shallow water, marine sponges can be collected by wading. They can frequently be found in tidepools and on the submerged surfaces of rocks. Many more sponges can be found by the collector swimming with mask and snorkel. He or she can examine the bases of wharves, pilings, and jetties and look for sponges living on the bottom in slightly deeper water. Of course people who use SCUBA gear can make collections over a much greater range of depths. They can also spend more time underwater studying their finds in the natural environment. Those people who are never able to get to the sea, because of where they live, can make very interesting collections of freshwater sponges from ponds and lakes.

After they have been collected, many sponges need merely to be cleaned and then allowed to dry. Better looking specimens can be made by soaking them in a 50 percent solution of Clorox bleach (half bleach and half water) for about twenty-four hours and then washing them off and putting them aside to dry. Sponges dry best in the open air in bright sunlight or in the kind of oven used to make pottery. Do not let your specimens become exposed to moisture while they are drying. That is, do not leave them exposed to rain or very humid weather. If a sponge is to be displayed in its natural condition, it should be preserved in a jar with 95 percent ethyl alcohol.

The sponges of the class Calcarea are usually quite small and they are best preserved in containers of alcohol. When alcohol is used, it should be changed once or twice before the container is stored in the collection. This is to insure that no dead organic matter is left to foul the clear alcohol.

Preservation in alcohol is also best for Demospongiae which have no spongin. On the other hand, the bath sponges, which of course possess a large amount of spongin, are very attractive when dried. The excavating demosponges usually have most of their "bodies" well hidden inside their chalky substrates. They can be preserved either by drying or storage in alcohol.

Other interesting demosponges to be added to a sponge collection are the suberitid sponges which have overgrown and dissolved snail shells occupied by hermit crabs. In fact, just such specimens (*Xestospongia halichondrioides*) and many others are being offered for sale by certain biological supply houses in Florida. Among the very beautiful objects for a collection are bits of dead coral or dead snail shells with sponges growing on them, or perhaps a preserved sponge crab with its miniature sponge garden on its back. Such objects are not uncommon in shallow warmer water like Florida and the Gulf of Mexico. They can be cleaned and dried like other sponges and

Left: A museum specimen of the bath sponge, Euspongia graminea *(about 4 x 11 inches)* Right: A museum specimen of the glass sponge, Venus's flower basket (Euplectella speciosissima), *showing the lovely lattice-like skeleton. Each unit measures about 16 inches in length.*

often make a very striking display.

The most beautiful of all sponges, when preserved, are the glass sponges of the class Hexactinellida. But these come from deep water and can only be obtained with the help of shrimpers or commercial fishermen. The beautiful Venus's flower basket, *Euplectella*, can be bought in curio or shell shops, and if one can be obtained with the two imprisoned shrimps inside, it is bound to be a real conversation piece.

After all the local species have been collected, you can increase your sponge collection by offering to exchange specimens with people living in other parts of the world. Sponges can also be obtained in exchange for other objects collectors might be interested in like seashells, starfish, and so on.

Remember that a good collection of local sponges with careful data will always be a valuable collection, especially for the department of invertebrates of any museum. Even if you plan to make a sponge collection only as a hobby, always try to be accurate and complete in your records. Take a field notebook with you on collecting trips and number the sponges you have collected each day. To make sure that you keep your specimens from becoming mixed up, use a plastic bag for each one. As you put the sponge in the bag, add a tag or label with a number on it. Enter that number in your field notebook and record data about the sponge. Your records should include the name of the collector, the locality, the depth of water collected in, the date of collection, the color in life, and any other information you think will be useful or interesting. After you preserve or dry your sponge, add a note to your records telling the date and type of preservation.

Precautions

Live sponges of all sorts—except the stinging sponges—are harmless when handled. But there is one precaution which should be taken, especially if you are handling sponges containing spicules. The spicules are tiny and sharp and almost invisible. Therefore it is wise to wash your hands carefully every time after you have held such a sponge. Otherwise you might absentmindedly rub your eyes with your hands and injure them with the spicules.

Identifying sponges

Sponges are not a very easy group to work with. They present some special difficulties to the person who wants to identify them. In the first chapter, we have seen that sponges come in many different colors and shapes. The same colors can be found in sponges of different species, and sponges of the same species can show some variation in color. It is also true that the shape of a sponge depends a great deal upon the spot where the sponge is growing or its habitat. The same species of sponge growing in undisturbed deep water may develop large branches, whereas in shallow water which is ruffled by wave action, the sponge will be merely a flat crust covering a rock. A sponge which is vase-shaped in deep water shows up as an irregular solid mass in shallow water or in rock crevices.

These variations make it difficult for people to identify many sponges on sight. Nevertheless, some sponges can be identified readily by their color, shape, or other external characteristics. Here are some steps to follow when you take a field trip to look for sponges. Always study your specimens carefully while they are still alive. You can carry a magnifying glass with you. Study the surface of the sponge while it is alive and under water. Look at the size of the oscula and the pores through which the water enters. Also look at the arrangement of these openings and note whether they are numerous or scarce. Look at the texture or consistency of the sponge. Of course you will not forget to make notes about the color and shape of the sponge. In fact, the more observations about sponges you make, the easier it will be to identify them.

If you cannot identify a sponge by its external characteristics alone, it may be necessary to study some internal characteristics. The form and type of spicules is the most important internal characteristic now used by spongiologists for identifying sponges. Fortunately it is possible to follow a very simple method and make temporary slides of spicules to study under the microscope. A tiny bit of the sponge is sliced off and placed on a microscope slide. A drop or two of Clorox is added. After the sponge cells have been dissolved and the spicules freed, a cover glass is added, and the spicules are ready for examination. The spicules on the slide can be compared with pictures of spicules in a sponge identification book. Making a permanent slide is a more complicated process and books on microtechniques should be consulted. (See Bibliography.)

EXPERIMENTS WITH SPONGES

It is almost impossible to keep sponges alive for a long time in a fair-sized tank or aquarium, even when aerated by a strong ventilating motor. As we have seen, a sponge is a kind of filtering machine which needs enormous quantities of fresh, food-laden water to stay alive. The average home tank cannot provide this. In Florida there are some saltwater swimming pools which are directly connected with the ocean tides. In such pools sponges can indeed live and flourish if they are allowed to.

However, even if sponges do not stay alive very long in a tank, they do stay alive long enough to perform some interesting experiments with them. If a small quantity of dye is dropped into a tank in which a sponge has been placed, one can quickly see how the dye is moved about by the strong volume of water pouring from the osculum. It is also possible to place particles of carmine in the tank and see them being dispersed through the osculum after they have been taken in by the sponge. You can make records of the time it takes these particles to pass through the water filtering system of the sponge.

Some zoologists perform a more delicate experiment. As we have seen, a sponge is made up of individual cells which, unlike those of the true metazoans, are more or less independent of each other. This is illustrated in the following experiment. A bit of live sponge is forced through some fine but strong cloth, such as heavy silk, and the separated sponge material is returned to the water upon a microscope slide resting on two small dishes in a fingerbowl of clean seawater. Care must be taken to separate the cells, not to mash up and destroy them. If one looks at the separated or disassociated cells under a microscope, it will be seen that it is mainly the choanocytes, the collar cells, and the amebocytes, the traveling cells, which have survived the rough treatment. In a short time the cells put out little extensions and begin to draw together. Soon a small piece of sponge body will have been formed. Eventually a new miniature sponge will result. This experiment has been performed many times, and when it is done with care and patience, it can turn out successfully.

There are many experiments and studies that you can make with sponges in their natural environments. Try some if you go on a vacation to the seashore. We will discuss ways of studying marine sponges since they are the most common, but freshwater sponges can also be studied in the same way.

Choose one or two common species and study their ecology. Here are some suggestions of things to look for. Make notes of where the sponges live with respect to the depth and the wave action of water, what animals live near them, whether they usually grow on a particular substrate, what animals you have seen feeding on them. Make a map of the locality showing where the sponges are found.

Take different substrates—clam, oyster, or snail shells, different types of rocks, and artificial substrates such as plastic and glass. Use a water-fast glue to attach these substrates to wooden boards. Hang the boards over the side of a pier, wharf, jetty, or

Sponge cells separated artificially and coming together again to form new sponges. Note the tailed choanocytes and the amebocytes.

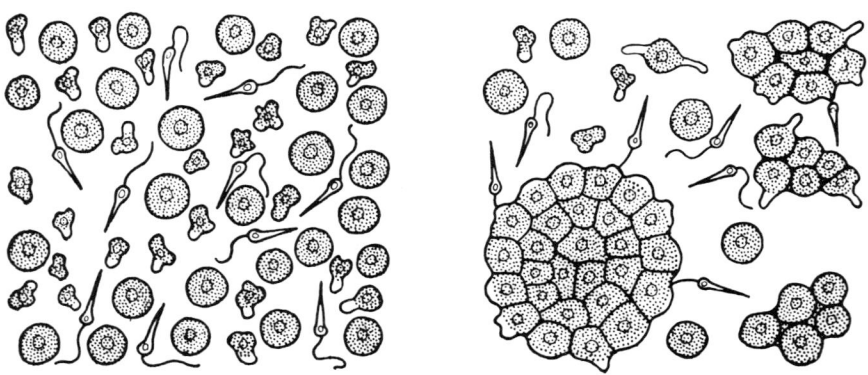

from some rocks in a protected area where sponges are commonly found. See which substrates sponge larvae settle on. You will most likely get larvae of many other marine invertebrates as a bonus. This in itself will be very interesting because it will allow you to make comparisons. You will also get an idea of the months when the larvae of different species begin to settle.

If your observations are thorough and your records are neatly and carefully kept, you might make some discoveries about the Porifera that have never been made before. In fact, what might have begun as an interesting hobby could become a life's vocation. Things like this have happened in the past.

SPONGES AND ZOOLOGISTS

It is true that in the past sponges have not been as widely studied as other creatures. Not many zoologists devoted all their time to them as others did to birds, mammals, fish, mollusks, and so on. This may be due to the fact that, with few exceptions, most sponges affect the lives of human beings very little.

But recently more and more scientists have begun to pay attention to the Porifera. Earlier we mentioned a vicious circle in connection with the weakened sponge fishing industry. Here we find something that is just the opposite: the more that sponges are studied, the more important facts might be discovered; and when more important facts are discovered, the more sponges will be studied.

The invention of SCUBA diving equipment and the increasing exploration of great ocean depths seem to have revived interest in the study of these relatively little-noticed creatures. As knowledge about them increases, valuable facts are being disclosed. The recent discovery of a possibly important antibiotic in sponges is a good example of this. More study will bring about the discovery of more important facts, important not only to biologists, but to mankind as a whole.

Glossary

Ageles (*a-jeel´-es*)—The elephant ear sponge

amebocytes (*a-meé-bo-sites*)—Mobile sponge cells which transfer oxygen and nourishment to other cells

archaeocytes (*ar´-key-o-sites*)—Sponge cells which play a role in reproduction. A kind of amebocyte.

asconoid (*as´-con-oid*)—The simplest kind of sponge body

asexual—Any process of reproduction which does not involve the union of male and female sex cells

bivalve—A class of mollusks whose members have two shells

Calcarea (*cal-caré-ee-a*)—The class of sponges which have limy or calcareous spicules. Also called Calcispongea.

calcareous (*cal-caré-ee-us*)—Composed of or containing calcium carbonate; limy or shelly

choanocytes (*ko-anń-o-sites*)—Sponge cells with a collar and flagellum; collar cells

cross-fertilization—The union of the egg cell from one individual with the sperm of another

Demospongiae (*dee-mo-spongé-ee-ee*)—The largest and most important class of sponges having a skeleton of glass spicules, spongin, a combination of both, or no skeleton at all; sometimes called the Demospongea

Euplectella (*you-pleck-tell´-a*)—The genus in which the glass sponge, Venus's flower basket, belongs

flagella (*fla-jell´-a;* singular, flagellum)—Little whips or cilia which drive the water current through the sponges

fertilization—The union of egg and sperm which begins the process leading to the development of an embryo gemmules (*jem´-youlz*)—Bodies formed in fresh water and some marine sponges by asexual reproduction and allowing many species to survive unfavorable conditions

genus (*jee´-nus*)—A group of related species which forms a subdivision of a family

Geodia (*jee-o´-dee-a*)—A genus of sponges, some of which look like pinkish cakes

Grantia (*gran´-tee-a*)—A genus of sponges named in honor of R.E. Grant, an early spongiologist

Gymnodinium (*jim-no-din´-ee-um*)—A protozoan that causes red tides when it is present in large numbers in marine waters

Hexactinellida (*hex-ak-ti-nell´-i-da*)—The class including the glass sponges with six-rayed, glassy spicules. Also called Hyalospongea.

invertebrate—Any animal without a dorsal or spinal column (backbone), such as a mollusk, starfish, insect, lobster, sponge, etc.

larva (*lar´-va*; plural, larvae)—The early stage of an animal and unlike the adult

leuconoid (*lew´-con-oid*)—The most complicated type of sponge body

limy—Consisting of or containing lime or limestone; *see* calcareous

luffa (*loo´-fa*; also spelled loofa, loofah)—A gourd which looks like a sponge; a false or vegetable sponge

megascleres (*meg´-a-skleers*)—The large spicules

Metazoa (*met-a-zo´-a*)—The subkingdom containing many-celled animals

microscleres (*mi-cro-skleers*)—The small spicules

mollusk—Any animal belonging to the phylum Mollusca, which includes snails, clams, squids, octopuses, etc.

myocytes (*my´-o-sites*)—Sponge cells which control the size of the oscula openings

organism—A single plant or animal

osculum (*oss´-kew-lum*; plural, oscula)—The opening in a sponge through which the water current escapes, bearing away waste products, and sperm is released

ovum (*o´-vum*; plural, ova)—An egg; the sex cell of a female

parasite (*par´-a-site*)—An organism that lives on or in another at the expense of its host

Parazoa (*pa-ra-zo´-a*)—The animal subkingdom to which the sponges belong

Petrosia (*pet-ros´-ee-a*)—A genus of sponges, some of which are maroon colored and platelike

phylum (*file´-um*)—The chief division in the classification of the Animal or Plant Kingdom; for example, the phylum Porifera pinacocytes (*pin-a-ko-sites*)—Sponge cells which make up the hard outer surface of many sponges

Porifera (*po-riff-er-a*)—The name of the phylum of sponges

porocytes (*po-ro-sites*)—Sponge cells which control the size of the pores

protoplasm—The basic living matter of all animal and vegetable cells

Protozoa (*pro-toe-zo-a*)—The subkingdom containing all one-celled animals

pseudopod (*soo-doe-pod*)—A temporary extension of protoplasm to serve as a foot, as in the amebas

regeneration—A process by which some lower invertebrates including sponges restore limbs or parts of the body which have been removed or otherwise lost

sessile (*sess-il*)—Permanently fixed; not free-moving; not mobile

scleroproteins (*skler-o-pro-teens*)—The "hard" proteins which include hair, fingernails, and spongin

Sclerospongiae (*skler-o-sponge-ee-ee*)—The class of sponges including the hard, corallike sponges

siliceous (*sigh-li-shus*)—Flinty, glassy, composed of silicon

sperm—The sex cell of the male

spicules (*spick-youl*)—The hard parts of the sponge skeleton made of chalky or glassy material. They may be separate or fused to form a network.

Spongidae (*sponge-i-dee*)—The family containing the commercial sponges

Spongillidae (*spon-jill-i-dee*)—The family of freshwater sponges

spongin—The sketetal material found in commercial and other sponges of the class Demospongiae

Suberitidae (*sub-er-i-ti-dee*)—A family of demosponges; the suberitid sponges

substrate (*sub-straight*)—The surface on which an animal lives; for example, rocks, sand, coral banks, etc.

syconoid (*sigh-con-oid*)—An intermediate type of sponge body

Terpios (*ter-pi-os*)—The sponge genus containing the "killer sponge" which is toxic to corals

Verongia (*veh-ron-jee-a*)—A genus of sponges, many of which are large and basket-shaped

zoanthid (*zoe-ann-thid*)—A relative of the corals and sea anemones often found growing on sponges

Annotated Bibliography

Barrett, J. H. and C. M. Yonge, 1958, *Collins Pocket Guide to the sea shore.* Collins, London.
    About a dozen British sponges are described and most are illustrated.

Buchsbaum, R., 1967, *Animals without Backbones,* second edition, University of Chicago Press, Chicago.
    Chapter 6 deals with sponges. Brief but clear treatment; well illustrated. A high school biology textbook.

Buchsbaum, R. and L. J. Milne, 1967, *The Lower Animals: Living Invertebrates of the World.* World Nature Series, Doubleday, Garden City.
    Devotes ten pages to sponges; discussion limited to the classes. Several excellent pictures of sponges.

de Laubenfels, M. W., 1932, *The marine and fresh water sponges of California.* Proceedings of the United States National Museum, vol. 81, 40p.
    An original paper which will help those interested in identifying sponges from the California area. Illustrations of spicules and whole sponges.

de Laubenfels, M. W., 1936, *Sponge fauna of the Dry Tortugas with material for a revision of the families and orders of the Porifera.* Tortugas Laboratories Publication # 30. Carnegie Institution of Washington, Washington, D.C.
    An important work that can be used to identify sponges from the Gulf of Mexico, Florida, and the Caribbean.

de Laubenfels, M. W., 1953, *A guide to the sponges of eastern North America.* Special Publication of the University of Miami, 32p.
    Can be used to identify sponges from the eastern coast of North America. Many figures of spicules.

de Laubenfels, M. W., 1955, *Porifera.* In: *Treatise on Invertebrate Paleontology,* part E, R. C. Moore (ed.), University of Kansas Press, pp. 21–122, figs. 14–89.
    The most complete, if quite technical, treatment of the phylum, down to genus. Well illustrated.

Hartman, W. D., 1958, *Natural history of the marine sponges of southern New England.* Peabody Museum of Natural History (Yale University), Bulletin 12, 155p.

    Drawings of spicules and photographs of whole sponges. Useful for sponge identification. Also gives information on reproduction.

Hartman, W. D., 1973, *Beneath Caribbean reefs.* (Publication of the Peabody Museum of Natural History, Yale University), vol. 9, pp. 13–26.

    An interesting and nontechnical account of Caribbean reefs seen by a spongiologist traveling to great depths in a mini research submarine. Photographs of coral reef inhabitants, including the sclerosponges.

Hyman, L. H., 1940, *The Invertebrates: Protozoa through Ctenophora,* vol. 1, McGraw-Hill, New York.

    An old but very complete work on the biology of sponges with an extensive bibliography and excellent figures. The chapter called "Retrospect" in volume 5 of the series gives a summary of investigations on sponges from 1938–1958.

Jewell, M., 1959, *Porifera.* In: *Ward and Whipple's Freshwater Biology,* W. T. Edmonson (ed.), second edition, John Wiley, New York.

    An illustrated guide to freshwater sponges of the United States.

Milne, L. and M. Milne, 1972, *Invertebrates of North America.* Doubleday, Garden City.

    Contains brief discussion of Porifera and a few excellent color photographs.

Pennak, R. W., 1953, *Freshwater Invertebrates of the United States.* Ronald Press Company, New York.

    Biology of the family Spongillidae. An illustrated guide to the identification of the North American freshwater sponges.

Ricketts, E. F. and J. Calvin, 1968, *Between Pacific Tides.* Fourth edition revised by J. W. Hedgpeth, Stanford University Press, Stanford.

    Many references to Pacific sponges.

Rozee, E. and L. Rozee, 1973, *Sponge Docks.* 510 Dodecanese Blvd., Tarpon Springs, Florida.

    An account of sponge fishing at Tarpon Springs, Florida, with many photos. Some inaccuracies in description of sponge biology, but gives interesting account of Greek sponge divers in America and the sponge industry in general.

Smith, R. I. (ed.), 1964, *Keys to marine invertebrates of the Woods Hole Region.* Contribution # 11, Systematics-Ecology program, Marine Biological Laboratory, Woods Hole, Mass.

    Chapter on sponges by Prof. W. D. Hartman, glossary, figures, annotated list of species. Good section on microtechniques.

Wells, M., 1970, *Lower Animals.* World University Library. Sponges in general are treated in a special section.

Zinn,, D. J., 1973, *A Handbook for beach strollers.* Marine Bulletin # 12 (Publication of the University of Rhode Island), 116p.

    Brief descriptions of a few sponges which occur in the northeastern part of North America. Illustrations.

Index

(Page numbers in **boldface** are those on which illustrations appear.)

Ageles, **10**, 11
amebocytes, 28, 29, 33, 69, **70**
Annandale, N., 36
archaeocytes, 29, 32
Ascon, 16
asconoid body type, 16, 21, **31**
asexual reproduction, 30

bends, 55
Bryan, P. G., 44
budding, 30

Calcarea, 21, 65
carrier cells, 33
Ceratoporella, 22
choanocytes, 27, **28**, 33, 69, **70**
cilia, 33
Cliona, 38–42, 46, 51
Cliona celata, 39
Cliona delitrix, **40**, 49
Cliona lampa, 41
Cliona laticavicola, **38**
Cliona schmidti, 12
collar cell, 25, 69
cross-fertilization, 33

Demospongiae, 11, 17–19, 21, 34, 36, 47, 52, 65
diver, 54
Dromia vulgaris, 47

ectyonin, 61
elephant's ear sponge, **10**, 52
Ellis, John, 9
Euplectella, 20, 66
Euplectella speciosissima, **66**
Euspongia graminea, **66**
Euspongia zimocca, **6**

flagella, 25, 27
flint, 62
free diving, 54

Gelliodes, **49**
gemmules, 31, 32, 37
Geodia, 11
glass sponges, 19
Goreau, T. F., 22, 46
Grant, R. E., 9, 10
Grantia, 10, **20**

77

gurry, 55
Gymnodinium, 57

Hartman, W. D., 22, 46
Hexactinellida, 20, 21
honeycomb sponge, 53
hookers, 53
Hyalonema sieboldii, **20**
Hyman, L. H., 30

Ircinia strobilina, 44

killer sponge, 44

Leuconia, 17
leuconoid body type, **16**, 17, 21
Leucosolenia, 21
Leucosolenia elenor, **31**
luffa (loofa), 51

megascleres, 15, 18, 21
Metazoa, 23, 24, 27, 31, 69
Microciona, **18**
microscleres, 15, 18, 21
Mycale, **48**
Mycale laevis, **45**, 46
myocytes, 29

Neofibularia nolitangere, 42
Neptune's goblet, **9**, 11

osculum, 25, 26, 30, 31, **48**, 69

Parazoa, 24, 31
Petrosia, 11
pinacocytes, 28, 50
plate cell, **28**
poison sponges, 51
pore cell, 27, **28**
Porifera, 8, 10, 12, 13, 15, 18, 19, 21, 24, 28, 31, 51, 64, 71
porocytes, 27, **28**
posca, 60
Poterion, **9**, 11

protoplasm, 23, 27
Protozoa, 23, 24
protozoan, 28, 57
pseudopods, 28

red tide, 57–59
regeneration, 31
Reiswig, H. M., 25, 35
reproduction, 30
Ruetzler, Klaus, 41
Russell, F. S., 12

Scapha, 21
scleroproteins, 17
Sclerospongiae, 21–22, 46
sexual reproduction, 30
silk sponge, 52, **53**
siphon-net sponge, 41, 42
Siphonodictyon, 41, 42
Siphonodictyon brevitubulatum, 45
spicules, 13, 15, 18, 21, 22, 29, 32, 47, 49, 50, 61–63, 67, 68
sponge crab, 47
Spongicola, 62
Spongidae, 11, 61
Spongilla, **37**
Spongillidae, 36, 38
spongin, 17, 18, 29, 32, 52, 65
Stelospongia, **19**
stinging sponges, 42, 51, 67
stinker sponge, 44
Suberitidae, 47
sulphur sponge, 39
Sycon, 17, 21
syconoid body type, 16, 21

Tedania ignis, 42
Terpios, 44
Tethya crypta, **50**
traveling cells, 69

velvet sponge, 53
Venus's flower basket, 19, 62, **66**
Verongia, 11, **25**, 37, **40**

Verongia archeri, **34**
Verongia gigantea, **26**
Verongia lacunosa, 12

wool sponge, 52

Xestospongia halicondrioides, 65
Xestospongia muta, 11

Yonge, C. M., 12

zoanthids, 49

Morris K. Jacobson, author of numerous scientific articles dealing with malacology and related subjects, is an associate in Malacology at The American Museum of Natural History in New York. Under a National Science Foundation grant awarded in 1969, he wrote several monographs on the land mollusks of Cuba. He is also coauthor of several popular books on mollusks, among them *Wonders of the World of Shells* in this series. Mr. Jacobson was trained in linguistics at Columbia University in New York and has taught foreign languages for many years. He is now retired from teaching and lives with his wife in Belle Harbor, New York.

Rosemary K. Pang received her B.S. degree in Biology from Tufts University and her Ph.D. from Yale. Currently a research associate in the Department of Biology, Brooklyn College of the City University of New York, she has published on the subject of excavating sponges from Discovery Bay, Jamaica, and described several new species. She and her husband, who is also a biologist, live with their two children in Belle Harbor, New York.